SpringerWien NewYork

Studies in Space Policy
Volume 1

Edited by the European Space Policy Institute
Director: Kai-Uwe Schrogl

Luca Codignola, Kai-Uwe Schrogl (eds.)
with Agnieszka Lukaszczyk and Nicolas Peter

Humans in Outer Space – Interdisciplinary Odysseys

SpringerWienNewYork

Luca Codignola
Kai-Uwe Schrogl
Agnieszka Lukaszczyk
Nicolas Peter

© 2009 Springer-Verlag/Wien

SpringerWienNewYork is a part of
Springer Science + Business Media
springer.at

Typesetting: Thomson Press (India) Ltd., Chennai

Cover: "Composite design courtesy European Science Foundation".

Printed on acid-free and chlorine-free bleached paper

SPIN: 12248469

With 47 Figures

Library of Congress Control Number: 2008935615

ISSN 1866-8305
ISBN 978-3-211-87464-6 SpringerWienNewYork

Preface

Humans and space

When faced with the issue of space exploration, one generally has an idea of the fields of study and disciplines that are involved: technology, physics and chemistry, robotics, astronomy and planetary science, space biology and medicine, disciplines which are usually referred to as the 'sciences'. In recent discussions, the human element of space exploration has attracted more and more the interest of the space sciences. As a consequence, adjacent disciplines have gained in relevance in space exploration and space research, in times when human space flights are almost part of everyday life. These disciplines include psychology and sociology, but also history, philosophy, anthropology, cultural studies, political sciences and law. The contribution of knowledge in these fields plays an important role in achieving the next generation of space exploration, where humans will resume exploring the Moon and, eventually, Mars, and where space tourism is beginning to be developed. With regard to technology, one might soon be prepared for this. Much less is this the case with space exploration by *humans*, rather than by robots. Robotic explorations to other planets across the solar system have developed in the past 50 years, since the beginning of the 'space age' with the presence of humans in nearby space and the landing on the Moon. Space exploration is now not only focused on technological achievements, as its development also has social, cultural and economic impacts. This makes human space exploration a topic to address in a cross-disciplinary manner.

Humanities research explores the origins and products of the human capacity for creativity and communication. Exploration is inherent to humans. Space exploration, and also human space exploration, as indicated above, has until recently mainly been dealt with by the sciences. Against this background, addressing the broader issue of humans in (outer) space with a focus on the human element and not only on technology imposes itself.

The European Science Foundation's Standing Committee for the Humanities (SCH) has taken a strong interest in the study of the implications of exploration by humans. This interest has led SCH to develop and lead an interdisciplinary initiative on this topic in close collaboration with the also ESF-based European Space Sciences Committee (ESSC).

The aim of this collaboration was to set up the first comprehensive and cross-disciplinary European dialogue on human space exploration and humans in outer space. Its aim was also to go beyond humans 'only' as tools in exploration, or as the better robot, and to address the inherent human quest for odysseys beyond the atmosphere. Aim was also to bring together scholars who usually have few reasons to meet in scientific forums, and exchange views in a non-traditional fashion.

Non-traditional because, beyond the technical aspects linked to human presence in space that have been studied by space scientists and engineers for the last five decades, humans in space pose challenges that go much further than the ability to survive.

On 22–23 March 2007, an ESF strategic workshop was organised at the University of Genoa (School of Letters and Philosophy, Department of Ancient and Medieval Studies) entitled *Humans in Space. A Humanities Assessment of the Implications of Space Sounding and Exploration*, addressing some of the issues identified above. Central theme was the role and situation of humans around the Earth, their place in exploration, and the search for life in the universe. Should humans explore space? Do the (cultural and economic) drivers for exploration require human participation? What are the human abilities and reasons to adapt to such extreme conditions as presented by the space environment beyond Earth? Are there scientific grounds that should lead man to be prepared for – ethical and societal – consequences of an encounter with extraterrestrial life? On the latter issue, reflecting on previous human encounters (cf. 1492) may help.

The cross-disciplinary interaction which resulted from this workshop paved the way for a conference on humans in outer space, organised on 11–12 October 2007 in Vienna, in collaboration with the European Space Policy Institute (ESPI) and the European Space Agency (ESA). The views and discussions presented at this conference are laid down in this volume. Scholars from a variety of disciplines and backgrounds, including history, cultural studies, religious studies, anthropology, the arts, policy, law, ethics and economics, but also technology, presented their views.

This resulted in a continued and further strengthening of the interdisciplinary European dialogue about human exploration of Moon and ultimately Mars, with a particular emphasis on the human element, as is illustrated by the contributions to this volume. The presentation discussions were structured around three odysseys in humans off the earth, as is also reflected in the structure of this volume. The conference has thus provided a unique European perspective by identifying various needs and interests of humanities and social sciences linked with space exploration.

From the Humanities, the conference has been a success. Not only on the scholarly level, through discussions with colleagues in other disciplines, with whom, indeed, regular interaction is not self-evident. The success has also been in demonstrating the necessity and productive contribution of humanities and social science disciplines understanding the universe in which we live, or will live in the future.

I would like to thank the participants to the workshop in Genoa, as well as the speakers at the conference in Vienna, for sharing their views in an open and cross-disciplinary manner. As one of the participants expressed it, the anthropologists and the rocket scientists finally talked to each other and more importantly: they also listened to each other.

Dr. Monique van Donzel
Head, Humanities Unit
European Science Foundation

Space and humans

What is exploration? Is it the pursuit of knowledge and science, wherever it leads us? It is that certainly, but that clearly is not the end of the story. Explorers throughout the ages have searched for fire, fresh water, food, milder weather, new hunting grounds, stone, minerals, spices, terra incognita, gold, precious stones, other life forms, rare animals, high mountains to climb, mysterious places to reach, and in the process bringing back answers, novel things to study, theories, and many more questions asked. Exploration seems to lie at the convergence of several drivers and behaviours, not necessarily compatible such as, curiosity (search for novelty and change); quest for new territories, conquests and riches; need to display and consolidate a nation's prestige. Thus exploration is not the realm of scientists alone: it is truly a societal enterprise that mandates defining and enforcing rules and ethics. Science seems to come out as a by-product of exploration, even if explorers were sometimes also scientists.

So what is exploration? Is it, in the words of modern explorer Mike Horn, to *"adapt to situations you did not plan for"*? Certainly, although I tend to prefer that famed replica from an equally famed television series: *"Exploration is to boldly go where no man has gone before"*. Space exploration certainly follows that definition. What could be bolder than for humans to sit on top of a largely untested and slowly exploding bomb, back in the early 1960s, if not the yearn to go where no one had gone before: around, and then beyond the limits of, the Earth itself? What could be bolder than to land a craft and a foot on the Moon, when nobody was certain that the ground would not collapse underneath? The rest of the story is known and largely deals with refining the science and technology that make these voyages possible. However, and from the very start, it was essentially that: for humans to go beyond the edge. It is thus quite paradoxical that space exploration remained for so long the remit of rocket scientists while for the general public, the human element was, and rightfully so, primordial.

Indeed can we leave it to machines to explore the universe in our place? Even though exploration will remain, for billions of human beings, a virtual adventure for a long time, possibly forever, it is difficult to relate to what a machine is doing 250 million kilometres away. Humans in space bring "un supplément d'âme" to exploration. Finally, since one of the ultimate quests of space exploration by humans and robots is to find whether or not we are alone in the universe, the search for extraterrestrial life is an extremely powerful driver: can we leave that to the robots? Naturally, there are places where humans can go, and places where only robots can work in. The exploration of the planets will continue to be done first robotically, and then with humans, but the key issue is that the debate "man or machine" is obsolete, and that humans should and will play a leading role in the exploration of space. Without it space exploration will simply lack an important societal and even scientific interest and perspective.

It is this realization, which provided the rationale for this ESF interdisciplinary initiative on Humans in Outer Space. It has been a very large success, bringing together colleagues from very remote disciplines who learned to talk together in the process, and it should also pave the way for new initiatives within the ESF and with the corresponding and very diverse scientific communities.

Dr. Jean-Claude Worms
Head, Space Sciences Unit
European Science Foundation

Joint acknowledgements

Our grateful thanks go towards all those, Steering Committee, workshop and conference participants, and ESPI staff, for making this initiative come to fruition, and to ESA and the Austrian Ministry for Transport, Innovation and Technology for supporting the Vienna conference. A particular word of thank you goes to Prof. Luca Codignola from the University of Genoa and SCH member and to Prof. Kai-Uwe Schrogl from ESPI, for their driving force in bringing the cross-disciplinary dialogue about, as well as to Ms. Marie Suchanova from ESF and Ms. Agnieszka Lukaszczyk from ESPI for taking care of the practical arrangements and organisation of the conference in a professional way.

M.v.D. and J.-C.W.

Table of contents

CHAPTER 5 Second Odyssey: Humans in space exploration: what effects will it have?

CHAPTER 6 Third Odyssey: Humans migrating the Earth: how will it affect human thought?

Introduction: towards a new vision for humans in outer space

Luca Codignola and Kai-Uwe Schrogl with Agnieszka Lukaszczyk and Nicolas Peter

Nothing has catalyzed a change in space policy on the global, the European as well as the national level more than the announcement of the U.S. President George W. Bush in January 2004 to launch a new U.S. Space Exploration Strategy.[1] It was not the 2004 Asian Tsunami, where Earth observation satellites were able to prove the essential contribution of space technology in saving lives and managing major natural disasters; it was neither the debate about the European satellite navigation system, Galileo, providing an immensely important strategic asset for European autonomy in a multitude of policy areas. No, it was the U.S. Space Exploration Strategy with its goal of bringing humans back to the Moon and further on step on Mars and beyond.

Moreover, this was not only a debate amongst policy makers, but it also reached the public and drew its attention in various ways. Breathtaking pictures of ice on Mars, a billion Internet hits during the Mars rovers' investigations, or the landscapes of the Saturn moon, Titan, where the European probe Huygens landed – they all demonstrate that humans long for knowing, what happens in our solar system and finally desire to go there. The public is also attentively following current human spaceflight to and from the International Space Station (ISS), where Europe has only very recently attached its own orbital research module.[2] The next public frenzy is already in the making: with upcoming commercial human suborbital flights, also known as space tourism. Hundreds of thousands of people have already expressed their interests to experience such short flights into outer space, and quite a few of them have already paid advances for the various ventures.[3]

Hence, humans in outer space are neither science fiction nor are they dull science. They are in the midst of policy debates and the public imagination. This debate, however, has been lead during the past decades with a rather narrow focus. Quarrels about budgets and fierce fights amongst scientific communities even lead to a general distrust of the European public in space programmes in the late 1980s and early 1990s. It was only in the late 1990s that governments and agencies began to understand the public interest and support in human space activities to be more openly reflected (and not only hidden) in their space endeavours and that they have to overcome the dichotomy between "utilitarian" and "trans-utilitarian" space activities.[4]

Using astronauts, as role models for engaging the youth in science in a more "aggressive" way is only one of the signals, topped only by the invitation during the German European Union Council Presidency in early 2007 to have an astronaut speak to the assembled heads of governments. Furthermore, an astronaut, Claudie Haigneré, recently became French minister for Research and another astronaut, Umberto Guidoni, became member of the European Parliament.

It is in this context that the European Science Foundation (ESF) initiated the process of bringing together the humanities and science to provide a new, broader look at humans in outer space. Prepared in the Genoa workshop of 22–23 March 2007, this led to the "Humans in Outer Space – Interdisciplinary Odysseys" conference on 11–12 October 2007 in Vienna. From the beginning the approach was non-traditional. Non-traditional because, beyond the technical aspects linked to human presence in space that have been studied by space scientists and engineers for the last five decades, humans in space pose challenges that go much further than their ability to survive, and questions that can only start to be addressed in the light of modern understanding of historical events. Thus, this approach went further than regarding humans only as tools for exploration or the better robots. It investigated the human quest for odysseys beyond the atmosphere as well as it reflected on the possibilities to find extraterrestrial life.

The Interdisciplinary Odyssey was organised along with space experts and scholars from the area of humanities as well as social sciences discussed the roles various disciplines such as law, philosophy, ethics, culture, art, psychology, increasingly will play in space exploration. The output of the conference was developed in form of the Vienna Vision on Human in Outer Space, which provides a unique European perspective in identifying various needs and interests of humanities and social sciences linked with space exploration. This book includes a selection of articles first presented at the Genoa workshop and later finalized for the Vienna conference, where the "Vienna Vision" was eventually formulated and approved. A few days before the conference took place, "Space Age" had reached its 50th anniversary[5], which allowed as well as asked for reflections on what has been accomplished until the present. In addition, that anniversary urged for creatively prospecting the future from various angles, since space activities and exploration are no longer focused on merely technological attempts. In fact, their development already have a great social, cultural and economic impact. Space activities are now entering an era where the contribution of the humanities – history, philosophy, anthropology, the arts as well as the social sciences, political science, economics and law – will become essential for the future of space exploration. Finally, the attentiveness for the societal complexity of activities in space is growing internationally.

The first part of the book provides a broad overview of the issue at stake through articles by Luca Codignola, Gerhard Haerandel, Thomas Ballhausen, Agnieszka Lukaszczyk, James Muldoon, Gísli Pálsson, Ulrike Landfester, Edi Keck, and Nicolas Peter. Varieties of different viewpoints build the foundation for the upcoming Odysseys. The scene is set through the discussions of the human being and its desire to explore the universe and to investigate what is really up there. The visions of the young generation for the future are explored. Moreover, the notion of "space" is discussed through its depiction in various forms of art as well advertising and marketing. All of this brings "space" closer to the reader and introduces him to the journey that follows.

The sessions of the conference were described as "odysseys" to signify that humankind is still on its way – or hasn't even started to leave. During the conference each odyssey was subdivided into four parts, which in detail illustrated the importance of each odyssey. The interdisciplinary approach, which was applied throughout the conference when examining the future of space exploration, contributed to a unique analysis in the articles provided by the authors. These will prompt many relevant questions while they shed a distinctive light on the topic of humans in outer space.

The First Odyssey evaluates the presence of humans in the Earth orbit and its consequences. Planet Earth is looked at from a different angle. It is treated as a home to all humanity; thus, it reinforces the need for care and protection of the planet. Claude Nicollier, Richard Tremayne-Smith, Gabriella Cortellessa and Frans G. von der Dunk contributed to this part. The question of identity is brought up, as once in space humans are most likely to identify with Earth as a whole instead of one's own country, region, etc. The idea of progress is discussed, as it is often associated with travelling to the Earth's orbit. It is also illustrated that the high technology and the need for innovation are often inspired by the human space flight. More than that, the various spin-offs are often beneficial for society at large and can facilitate further research, thus, promoting science. In addition to the technological progress, social progress is discussed. It is emphasised that space exploration offers many possibilities for international co-operation through en-deavours such as ISS. Mutual understanding and teamwork are crucial for successful space mission and could perhaps promote further collaboration on Earth.

Technology in itself is an important part of the First Odyssey, as humans continue to increasingly depend on various technological advancements. This brings up the issue of human–machine relationship and how it may evolve over time. The legal aspect of human space flight has not been neglected. The importance of law is discussed, as with the further space exploration the need for development of the space law will become necessary in order to cultivate the peaceful uses of outer space. Moreover, human rights are also considered, as perhaps new moral challenges may face humanity in the future.

The Second Odyssey focuses on the various effects space exploration may have on humans. Wolfgang Baumjohann, Jacques Arnould, Stephan Lingner and Ulrike Bohlmann offer a great illustration of the subject matter. They argue that the human factor is essential in space exploration. It not only fosters the collaboration amongst societies and cultures but it also opens up the opportunity to follow the thrust of scientific and cultural curiosity. The need for discovery has been an essential part of human evolution and is precisely what continues to inspire humans to explore new places and search for new worlds.

The authors have recognized that human space flight would very likely have an effect on the various aspects of culture. Through regional co-operation European values and priorities may be redefined and perhaps be strengthened. The generation which grew up in the space era has already had a different outlook on the world, for instance on environmental issue, in comparison to the past generations.

The Third Odyssey is more of a philosophical nature. It concerns humans leaving the Earth permanently; hence, migrating to distant planets. The issue of how would that would affect human thought is posed. Contributions by Gerda Horneck, Paolo Musso and Debbora Battaglia make for a very stimulating read. The idea of habitat is examined owing to the fact that once leaving the Earth humans would have to establish settlements elsewhere. First children would be born in space. True space generation would be established. These experiences would require adaptation to the new environment as life would most likely differ very much from the one on Earth. While considering these issues, some authors felt it was important to reflect on the issue of belief systems. The environment humans are surrounded with often alters the scheme of faith, religion, morals, and values. Would some of these systems collapse if humans leave the Earth permanently? Would new systems of beliefs develop? The answers to these questions and more are discussed throughout the book. The discussion is further instigated by the thought of possible encounters with other forms of life in outer space. A new chapter in the human history would begin should humans discover they are not alone in the universe.

This book opens a door to a very much-needed dialogue concerning human space flight in a variety of disciplines. Such dialogue is necessary in order to make societies aware that space exploration involves much more than high technology and science in order to be successful. Elements of what is needed are contained in the conference results, described as "The Vienna Vision on Humans in Outer Space". The Vienna Vision provides the context as well as the main findings that this interdisciplinary quest has produced. It is addressed to the programme- and decision-makers for reflecting on so far unnoticed or undervalued arguments and reasons for a human presence in outer space. The Vienna Vision has already been brought to the attention of its communities.[6] Very rarely was an initially academic venture been able to address its finding so directly. This one was able to do.

The editors are extremely glad ESF has taken the initiative to let the ideas of an illustrious group of scientists and practitioners representing such a wealth of disciplines and communities to enter the odysseys, which reached its Ithaka in the form of the Vienna Vision and this book. The venture of humankind to broaden its presence in outer space has only begun. Next year, the 40th anniversary of the first human landing on the Moon will be celebrated.[7] We hope that this book will provide inspiration and assistance to scope the future ahead.

---

[1] White House Official Website. "President Bush Announces New Vision for Space Exploration Program" 27 February 2008 http://www.whitehouse.gov/news/releases/2004/01/20040114-1.html
[2] Universe Today. "Columbus Module Attached to ISS after Eight Hour Spacewalk" 11 February 2008 http://www.universetoday.com/2008/02/11/columbus-module-attached-to-iss-after-sts-122-spacewalk
[3] Space Adventures Official Website. "More Space Flight Experiences" 27 February 2008 http://www.spaceadventures.com/index.cfm?fuseaction = Other_Spaceflight_Experiences.welcome
[4] Gethmann, Carl Friedrich. "Manned space travel as a cultural mission" Poiesis & Praxis, 4 Dec. (2006): 239–252.
Schrogl, Kai-Uwe, Rohner, Nicola and Lingner, Stephen. "A New Approach in Justifying Space Activities – Overcoming the Dichotomy of Utilitarian vs. Trans-utilitarian" 2nd Space and Society Conference, March 2007 ESA/ESTEC, Noordwijk, Netherlands.
[5] On 4 October 1958, Sputnik had been launched.
[6] By Nicolas Peter and Agnieszka Lukaszczyk at the 9th International Lunar Exploration Working Group's (ILEWG) International Conference on Exploration and Utilisation of the Moon (ICEUM9/ILC2007) in Sorrento, Italy on 24 October 2007 and by Jean-Claude Worms and Gerhard Haerendel at the International Space Exploration Conference co-organised by ESA and DLR in Berlin on 8–9 November 2007 where about 300 space policy stakeholders, including head of space agencies gathered. In addition it was brought to the attention of the United Nations Committee on the Peaceful Uses of Outer Space (UNCOPUOS) Scientific and Technical Subcommittee on 15 February 2008 by Kai-Uwe Schrogl.
[7] On 21 July 1969, Neil Armstrong and Buzz Aldrin set foot on the Moon.

List of acronyms

A
ACHME: Advisory Committee for Human Spaceflight, Microgravity and Exploration Programmes
AI: Artificial Intelligence
AMOCT: Advanced Mission Operations Concepts and Technologies
APSI: Advanced Planning and Scheduling Initiative
ASI: Italian Space Agency (Agenzia Spaziale Italiana)
AU: Astronomical Unit

B
BMVIT: Austrian Federal Ministry for Transport, Innovation and Technology (Bundesministerium für Verkehr, Innovation und Technologie)

C
COSPAR: Committee on Space Research
CSA: Canadian Space Agency

D
DARA: former German Space Agency (Deutsche Agentur für Raumfahrtangelegenheiten)
DLR: German Aerospace Center (Deutsches Zentrum für Luft- und Raumfahrt)
DNA: Deoxyribonucleic acid

E
EANA: European Exo/Astrobiology Network Association
ECGS: European Center for Geodynamics and Seismology
ESA: European Space Agency
ESF: European Science Foundation
ESI: European Standardisation Institute
ESO: European Southern Observatory
ESOC: European Space Operations Centre
ESPI: European Space Policy Institute
ESSC: European Space Sciences Committee
ET: Extra-Terrestrial

EURECA: European Retrievable Carrier
EVA: Extra-Vehicular Activity

F
FP 7: 7th EU Framework Program
FPSPACE: Friends and Partners in Space

G
GEO: Geostationary Orbit
GWU: George Washington University

H
HRSC: High Resolution Stereo Camera
HST: Hubble Space Telescope

I
IAA: International Academy of Astronautics
IAC: International Astronautical Congress
ICSU: International Council for Science
IGA: Inter-Governmental Agreement
IISL: International Institute of Space Law
IPCC: Intergovernmental Panel on Climate Change
IR: Infrared
ISS: International Space Station

J
JAXA: Japan Aerospace Exploration Agency

L
LDEF: Long-duration Exposure Facility
LEO: Low Earth Orbit

M
MARSIS: Mars Advanced Radar for Surface and Ionosphere Sounding
MIT: Massachusetts Institute of Technology
MOL: Manned Orbiting Laboratory
MPE: Max-Planck-Institut für extraterrestrische Physik

N
NASA: National Aeronautics and Space Administration
NEO: near-Earth-object

NGO: non-governmental organisation
NPS: Nuclear Power Source

O
OMEGA: Observatoire pour la Mineralogie, l'Eau, les Glaces et l' Activité
OST: Outer Space Treaty

P
PET: Positron Emission Tomography
POR: Payload Operation Request

R
RMS: Remote Manipulator System

S
SACSO: Safety Critical Software
SAIL: Shuttle Avionics Integration Laboratory
SARS: Severe Acute Respiratory Syndrome
SETI: Searching for Extra-Terrestrial Intelligence
SFX: Science Fiction
SGAC: Space Generation Advisory Council
SPIDER: Space-based Information for Disaster Management
and Emergency Response
SSMM: Solid State Mass Memory

T
TSS: Tethered Satellite System

U
UFO: Unidentified Flying Object
UK: United Kingdom
UN: United Nations
UNCOPUS: United Nations Committee on the Peaceful Uses of Outer Space
UNESCO: United Nations Educational, Scientific and Cultural Organization
UNOOSA: United Nations Office for Outer Space Affairs
U.S.: United States

W
WIPO: World Intellectual Property Organisation

List of figures and tables

Chapter 1 Setting the scene

Chapter 2 Can we compare?

Chapter 3 "Spatiality" – Space as a source of inspiration

Chapter 4 First Odyssey: Humans in Earth Orbit: what effect does it have?

Chapter 5 Second Odyssey: Humans in space exploration: what effects will it have ?

Chapter 6 Third Odyssey: Humans migrating the Earth: How will it affect human thought?

Chapter 7 The Vienna Vision on Humans in Outer Space

CHAPTER 1

SETTING THE SCENE

1.1 Summary

Luca Codignola

Launched in August 2007, the Phoenix Mars Mission of the North American Space Agency (NASA) finally tasted a measure of success when, on 6 June 2008, its Mars Lander made its first dig into Martian soil and its Robotic Arm scooped up a certain amount of reddish material from the top four centimeters of the planet's surface. The mission, NASA's first in its Scout Program, was designed to verify traces of volatile element, water in particular, and, consequently, to assess Martian habitability potential. The programme has been some years in the making. It is almost a year since the rocket was launched. Furthermore, the first Moon landing took place in 1969. That was more than a generation ago. Are there alien entities somewhere in the universe eagerly and impatiently waiting for us to arrive? Is humankind moving too slowly in space?

Consider this. According to Gerhard Haerendel, one of the contributors to this chapter and a hard scientist himself (a physicist by trade), the solar system was born 4.6 billion years ago, the oldest known rocks were formed 4.4 billion years ago, and the first traces of microbial life are 3.5 billion years old. Furthermore, it took some three billion years to develop on Earth multi-cellular life – from which man and woman took their first breath of life. In this chronological framework surely we can wait another few years before we return to the Moon *en masse*, or we establish our first space station around Mars, or we send a lander to Europa. This is Jupiter's sixth satellite, a crusty sphere perhaps consisting of ice, an element which implies water, and, yes, habitability.

In spite of having made space his profession, Dr. Haerendel, however, is more inclined to keep us down to earth and dampen our amateurish enthusiasms. José Gabriel Fumes, a Jesuit Catholic theologian and astrophysicist who since 2006 heads the Vatican Observatory, does not consider the existence of extraterrestrial life as in contradiction with the Christian ideas of creation, incarnation, and human redemption, implying that there might well be other forms of life somewhere in outer space. How could God's omnipotence be limited to such a small planet within the magnitude of the universe? Dr. Haerendel does not deny such a possibility, but well shows how difficult it is, and indeed improbable, that humankind will ever find out, stuck as we are with our limited solar system and even more so with the minuscule lifespan granted to each of us.

3

Hard facts, however, are not the only nourishment of a human being. When Christopher Columbus left for the New World, he knew what was pushing him: "To increase the Holy Christian Religion . . . to seek gold and spices and to explore land" (*Journal*, 6 November 1492). Only the second element of the classic triad has to do with hard facts. In today's terms, these may translate into a number of factors that would enrich not only some daring entrepreneurs, but also the whole humankind. This may happen through new sources of energy, new room for an exploding population, new cures for old diseases through a better understanding of human physiology, the monitoring and avoiding of natural disasters, and the betterment of developing countries through the spreading of new technologies. Agnieszka Lukaszczyk's article, reporting on the extensive survey mapping young people's visions for the next 50 years of space exploration, makes it clear that the generation that will produce the leaders of the future expect nothing less from projects such as Phoenix Mars Mission. Their utilitarian approach, however, is not devoid of a certain optimism with regard to human beings' willingness and ability to co-operate, let alone of an ethical preoccupation with the spiritual improvement of humankind. In a way, this ethical preoccupation is the current translation of Columbus's desire to "increase" the Christian religion, the first element of his triad.

As for Columbus's third and final element, curiosity, this is, perhaps, the most overwhelmingly present, albeit the most impalpable one. Dr. Haerendel describes it as "the enormous philosophical relevance of the question of whether we are or are not alone in the universe." Curiosity of is often more visible in fiction than in real projects taking place in actual life. For many years, actually since the invention of the moving pictures depicting fictional events, we have observed progenitors of the Mars landers descending upon the most distant corners of the universe; spacecrafts orbiting the Moon; astronauts disappearing into hyperspace; and alien entities – mostly ugly, dangerous, and lethal – battling civilization and its superior moral principles. Thomas Ballhausen surveys the history of science fiction through a selection of 90 moving pictures, from *Voyage dans la Lune* (1902) to *Sunshine* (2007). His list is far from exhaustive, witness the fact that he has elected to leave out classics such as *Close Encounters of the Third Kind* (1977) and two great franchises such as *Star Wars* (1977–2005) and *Star Trek* (1979–2002 so far). However he who is that curiosity is never absent from the genre, in that future scenarios are imagined that depict new technology, new governance, new political systems, new societal aggregations, new "encounters" (on which more later), let alone utopian visions of human progress. It is Ballhausen's contention that science fiction in moving pictures has always responded to actual political situations, but has also been affected by real space programmes. For example, the Cold War produced the best decade of science fiction moving pictures ever. One could also add that the Apollo programme was another turning point in the genre.

What, then, about future encounters with alien beings? Fiction has long imagined such an occurrence, but, as we know all too well, there is no accepted evidence of such encounters having happened in the past. Furthermore, most scientists who actually work on space programmes (such as Dr. Haerendel), are rather skeptic and only give some chances of relative success to programmes such as Searching for Extra-Terrestrial Intelligence (SETI). Historians A.E. (Alfred Worchester) Crosby, Jr., and Luca Codignola, whose articles we have left for last in this introduction, allow that they are not equipped to predict future encounters. However, they emphasize that encounters between different entities have indeed happened in the past on Earth itself. Furthermore, such encounters have not been inconsequential.

Dr. Crosby, one might recall, is the one who through his 1972 book imposed the notion of "Columbian exchange" on the community of historians and made it a commonplace locution even among lay readers. According to him, it was the biological exchange who wiped out a vast proportion of the aboriginal peoples of the Americas and made Europeans suffer the same fate in Africa. The same biological pattern, Dr. Crosby shows, was repeated over and over in other regions of the world, such as the Pacific Islands. In the article he has written for this book, Dr. Crosby equates the case of Hawai'i, given its position in the centre of the Pacific Ocean, to that of a distant planet positioned somewhere in the universe. The catastrophic consequences that devastated the Hawai'ian human beings once encounter took place, let alone its flora and fauna, are a dire warning to what may happen when encounters take place in outer space. Signs of it, Dr. Crosby maintains, are available aplenty. We humans, he writes, "are miniature ambulatory jungles."

For his part, Codignola mostly agrees with Dr. Crosby with regard to the nature of the biological exchange. He points out, however, that biological exchange also took place alongside ideological exchange. Codignola adds that, from the Western point of view, ideological exchange took place within the ideological framework of the Christian church. There the real trauma was represented by the arrival of the Mongols into the world scene in 1221. Although the Columbian voyage is normally mentioned as the quintessential encounter between two worlds, the discovery of America then represented a comparatively minor psychological trauma. Furthermore, in picturing future scenarios, Dr. Crosby seems convinced that the meeting of foreign worlds cannot bring but catastrophes, as it happened in the past. Codignola is more skeptic, in that he believes that nothing, really, can prepare humankind for an encounter of this kind, because we only know one side of it and variables are as innumerable as they are unknowable.

1.2   Micro-organisms and extraterrestrial travel

Alfred W. Crosby

I am taking as given that more practical means of space travel than exist now will be developed in the next century or two and that colonization – actual settlement – of extraterrestrial bodies will follow. Neither of these is certain or even close to certain, but they might happen, and if they do, problems will arise about which we had better start thinking beforehand, not during. We usually approach with a technological bias the subjects of extraterrestrial travel and possible colonization. When destinations are beyond our planet, the problem of simply getting from here to there is so intimidating that we initially think of the challenge in terms of engineering. But we may discover that the tougher problems may be biological. In our next decade spacecraft will be returning with extraterrestrial samples and more probes will be launched to get more samples than ever before. Right now we need to think at least as carefully about the life forms that may be on board space vehicles as we do about the vehicles per se. What will the organisms we will carry with us outbound do in their new environments? Also, what about the possibility that there are organisms native to the planets, asteroids, etc., to which we will travel and from which we may bring alien organisms to Earth? Will they survive here at all? Will they thrive here to our advantage or disadvantage, i.e., to the detriment of our health, food sources, and earthly ecosystems in general?

This all may sound like science fiction. However, I am not inspired by fiction but by history, by what happened when people first made a habit of crossing the great oceans, i.e., *post* 1492, establishing commercial and political and therefore biological connections between the continents and with previously remote islands. These lands were all chockfull of plants, animals, and microlife for enriching human visitors and for infecting them with unfamiliar diseases. The latter phenomenon still continues. I point to the spell out acronym (SARS) scare of a few years back and to the current threat of avian influenza. If today east Asia can produce pathogens to threaten Europe with pandemics, what might Jupiter's moon, Europa – just possibly juicy with life – provide for our entertainment? One might say that examples of what human ocean-crossers triggered by carrying micro-organisms with them are of no use to us in our considerations of the consequences of space travel. Such as we might call Columbian examples are too simple, too slow in pace, and too slight in magnitude – entirely too *earthly* – to help us think clearly about interplanetary contacts. I answer that, on the contrary, these examples can be very helpful in leading us to good questions, which are the prerequisites of good answers.

Let us look, for example, at the Hawaiian islands, which are today the crossroads of the Pacific world, but a few human generations ago were biologically and anthropologically what they still are geographically, i.e., the most remote and isolated major archipelago on the planet. Looking to Hawaii for hints about the possibilities and implications of space travel may strike the reader as absurd. Hawaii is green and hospitable and Mars, to choose what is likely to be the first planet humans will actually visit, is grey and inhospitable. A comparison of the two seems a waste of time, but we must not let ourselves be misled by technicolor contrasts. The pertinently significant contrast between Hawaii and Mars is not green vs. grey, but the presence of life here on Earth compared with the possible presence of any life at all on Mars. Mars may not be totally barren. If there is life in any form there, our concept of the universe and of ourselves will change instantly and massively and we will be forced to cope with opportunities and dangers of a magnitude undreamed of in Christopher Columbus's era.

The Hawaiian archipelago is geographically remote, not 100 million kilometers away, but 3000 or so away from the nearest continent and several hundred from the nearest islands capable of supporting more than a handful of humans. It is also remote in its biota. Ninety-six per cent of its native flowering plants occur naturally nowhere else. When Europeans first arrived – Capt. James Cook and his sailors in 1778 – the only mammals there were the dog, pig, and rat, all brought in by the Hawaiians, and, of course, the Hawaiians themselves, mammalian every one. The only truly native mammal of the Hawaiian islands is a bat. In 1778 William Bligh, one of Cook's captains, estimated the Hawaiian population at 242,000.[8] We get our first dependable population counts from the Yankee missionaries who started arriving in the 1820s. These *haoles* (Hawaiian for outsiders, usually white folks) believed in arithmetic, counted and calculated seriously, not artistically, and stayed on and spent the rest of their lives on the islands. By the 1840s Hawaiian population statistics were among the most respectable in the world. In 1823 the missionary-demographers estimated the Hawaiian population as 135,000; in 1850, as 84,000. By 1878 there were only 48,000 Hawaiians, even including some of only partly Hawaiian ancestry. That marks a drop of at least four-fifths in a century.

What explains this tragedy? Genocide? The Hawaiians were ruled by their own royalty until the latter years of the 19th century and while there were murders and at least one massacre, there was no slaughter to compare to what happened on the American mainland. Slavery? Nothing to be compared to what happened in Africa and America. Emigration? Some men left as sailors, but not many relative to the total population, and very few women left. Gross exploitation of laborers on plantations à-la-West Indies to cultivate products for export? Indeed there was some of that, but later in the century after the worst years of population decline. Cultural dislocation? There must have been plenty, but I doubt that many people die because they are confused and depressed, though there was a decline in the birth rate and may have been increased infanticide.

Infectious disease? After 1778 the Hawaiians were in ever increasing contact with people from lands with big human populations, particularly dense in the ports, and large herds of livestock with which humans exchanged diseases. To cite a specific example of what could happen, take the case of King Liholiho and Queen Kamamalu, who sailed to Britain early in the 19th century. Both died in London in 1824 of measles, a disease yet to debark in their islands. Early in the 19th century, 1804 perhaps, there was an epidemic of the semi-mythic *oku'u*, about which science knows nothing, but which Hawaiian tradition credits as being particularly fatal. In 1824 three epidemics swept the islands – measles, whooping cough, and influenza. The death rate from all three added together was estimated at 10%. In 1853 the most dreadful killer of American Indians, smallpox, arrived. The death rate that year was 105 per 1000.

These assaults on the Hawaiian population were accompanied by a steady drumbeat of a new and now constant factor in the islands' demography: sexually transmitted disease (STD). Hawaiians, who may have had little experience with STDs before 1778, found themselves subject to the unwavering attentions of tens of thousands of sexually hungry men – sailors. The Hawaiians suffered from venereal infections, which not only killed many, but must have sterilized a lot of the women. The Hawaiians' population dropped from no less than 242,000 to 48,000 in one century. Similar death rates afflicted many newly contacted populations – Aztec, Incan, Maori, the indigenes of Siberia – and, I should note, among Europeans as well when they landed on the malarial shores of tropical Africa. Such abrupt explosions and implosions of invading and indigenous organisms may occur on planets and other heavenly bodies we land upon. Maybe there is no life out there to endanger visitors or to be endangered by visitors, but surely we should restrain ourselves from leaping to that conclusion. We have a profound duty to act as if there were life on or in Mars and elsewhere out there.

The astrobiologists divide the subject of exchange of organisms between heavenly bodies into two categories, "forward contamination" and "back contamination." The first deals with the possible effects of our bringing earthly organisms to extraterrestrial destinations, infecting them with immigrant organisms, possibly disrupting whole ecosystems, and most certainly erasing much of their value to us for providing evidence of how life evolved wherever found.

Intellectually, our chief problem in thinking about such matters is that thus far we have only one example, our earthly one, of the origin and evolution of life. We are intellectually hobbled because we normally reason about complicated matters by making comparisons and ferreting out correlations, but we cannot compare X to Y if we only have X. We are obliged to be very careful about Y, in this case extraterrestrial life, if we find any.

The second kind of contamination, "back contamination," refers to the possible effects of bringing extraterrestrial organisms to the home planet, to Earth. The organisms we are most worried about carrying back – and, for that matter, forward – are those which are the best adapted for easy pick-up and delivery, the micro-organisms. Their threat demands the strictest standards and sophisticated planning. Such can lead to silliness, of course. Some officials in the United States' National Aeronautics and Space Administration (NASA) want the robot vehicles inspecting Mars to avoid the gullies because that is where water may have flowed and where invading micro-organisms from Earth carried by imperfectly sterilized machines would be most likely to establish themselves, disrupting our investigations. Other NASA officials fume and point out that in large part we are investigating Mars precisely to find if there is any life there, and the likeliest location for native Martian life is the gullies.

The Americans in NASA and, presumably, equivalent Russians and others with inclinations for space travel, have taken the standard earthly public health precautions against micro-organism stowaways; for instance, meticulous medical examinations of the human travellers before and after missions. German measles among NASA personnel almost led to the cancelling of an early Moon visit. Non-human living travelers (mice, bacterial samples, etc.) and, of course, of all materials collected from extraterrestrial bodies, are sterilized or kept in isolation. The first teams of humans returning from the Moon were strictly quarantined for days. And, of course, everything else that made the roundtrip – the mice, the equipment, and entire spacecrafts, etc., – have been sterilized. Even so, there is still plenty about travelling to and returning from other planets, comets, and what-have-you to worry about. Just because the Moon is encouragingly dead does not mean that other bodies beyond our atmosphere are, too. There are no gullies on the Moon, but there are on Mars. Perfect quarantine is impossible in practice. For example, in 1969 the Apollo 12 mission visited the Moon and returned to Earth. One of the items its astronauts brought back home was a camera, part of a Surveyor probe deposited on the Moon 3 years before. Back on Earth, micro-life, specifically streptococci, were discovered inside its camera, healthy and capable of growth. Twenty-two years later Pete Conrad, commander of Apollo 12, said: "I always thought the most significant thing that we ever found on the whole . . . Moon was that little bacteria who came back and lived and nobody ever said [anything] about it".

Two explanations for the streptococci being in the wrong place at the wrong time have been offered. The first was that the microbes were earthly passengers from the Surveyor probe which had somehow survived years on the Moon despite the temperature extremes, desiccation, and radiation. If true, that would indicate that microbes were better suited for space travel that we had believed, a disturbing thought. Less disturbing was the official explanation that the streptococci in question had really never left the Earth, but were the product of a "breach of sterile procedure". One of the instruments being used to scrape samples off the Surveyor probe for culturing after the Apollo 12 astronauts brought it back home had been set down on a non-sterile lab bench where it must have acquired the streptococci. It was then used again, depositing the streptococci on the probe. Said microbes had not travelled millions of kilometers but only a meter or two. Absolute sterilization is hard to maintain in practice, and the Apollo 12 astronauts and their aides were thoroughly human and therefore prone to error (Apollo). We humans are inclined to think of the Earth in terms of humans and most certainly of multi-cellular life. The truth is that it is the bacteria planet. Here life in one-cell packages encompasses over 90% of genetic diversity and an overwhelming majority of the total biomass. Single-celled life forms are everywhere, even on the exteriors and in the interiors of the most neurotic hand-washer among us. We humans are miniature ambulatory jungles whereon and wherein millions upon millions of micro-organisms dwell, which we carry with us everywhere we go. We are in and of ourselves excellent means for exchanging microlife between heavenly bodies.

Let us consider the probable characteristics of the life forms that may exist on the moon, Mars, the other planets, their moons, the asteroids, and which may be somehow surviving in space dust. When our rockets first began to fly beyond our atmosphere, our assessment of the possibilities of possible manifestations of life was limited by what we knew of life at the time. That was derived from what we knew of earthly life, all of which, as far as we knew, lived on or close to the planet's surface in environments that even at their worst did not exceed the limits of what scientific common sense imagined as tolerable. Then came the debut of the well-named *extremophiles,* earthly organisms – almost all of them single-celled – that exist and even propagate in environments that would kill you and your piggy-backing microbes instantly. We have found the extremophiles in the depths of Yellowstone's boiling pools; alongside tiny volcanoes kilometers deep in the oceans; and in ice that has not been liquid for millennia; at incredibly high altitudes; luxuriating in liquids of extreme acidity and alkalinity; in crustal rock; at temperatures that we used to think prohibited life; and at pressures a thousand times greater than we experience personally. Hence the name extremophile.

The first extremophiles to be sequenced were the *Methanococcus jannaschii*, single-celled microbes living near hydrothermal vents 2600 m below the surface of the sea, where temperatures approach the boiling point of water, and the pressure is sufficient to crush an ordinary submarine. There, the *Methanococcus jannaschii* survive on carbon dioxide, hydrogen and a few mineral salts. In fact, it cannot tolerate oxygen. An even more celebrated extremophile lives in Antarctica under ice four-kilometer thick over a reservoir of fresh water the Russians named Lake Vostok. They drilled a core vertically though that ice, 3623 m in length, the longest yet. There they wisely stopped for fear that they might penetrate all the way to the liquid water at the bottom and contaminate it. Toward the far end of the core, they found a community of living microbes. U.S. microbiologist David Karl, who has worked on the Vostok extremophiles, has stated: "Our results extend the possible limits of life on Earth and elsewhere in the Universe" (Britt). The Vostok extremophiles have inspired new interpretations of old data, and in January 2007 give first name and middle initial if any Schulze-Makuch of Washington State University and give first name and middle initial if any Hootkooper of Justus-Liebig University speculated that we actually found life on Mars way back in the 1970s, but did not recognize it because we did not know about extremophiles then, that is, we did not know that there might be organisms that rely on hydrogen peroxide in their metabolism (Washington). To utilize a tired American cliché, you do not have to be a rocket scientist to wonder if organisms as tough as the extremophiles – some benign, some possibly not – might be waiting for us on Mars and the extraterrestrial bodies elsewhere, if not on their exteriors then in their interiors. Could Martian extremophiles live here? That is unlikely and even if they did establish a beachhead on Earth, they probably would not endanger us. After all, they would be the products of a stream of evolution separate from ours and would have had no opportunity to adapt to our bodies and behaviors. On the other hand, they just might adapt fast, in which case the magnitude of the possible damage they might cause would be unpredictable. They might endanger not merely individual health, but the function and balance of major ecosystems.

Let us now consider what I would describe as Darwinian Probabilities. Space colonization, a possible (but by no means certain) sequel to space travel, would have unexpected side effects. Space colonization would settle human beings in separate homelands whose citizens would rarely meet, much less interbreed, homelands with different strengths of gravity, different kinds and degrees of radiation, different kinds of atmosphere, different life experiences for their inhabitants – differences which would spur divergent evolution. For thousands of years before Columbus, Ferdinand Magellan, Cook and other European ocean-spanners, we humans lived in a geographical arrangement a bit like the above, i.e., separate homelands with different environments. Our Pleistocene ancestors had migrated out of Africa to the far corners of the world. There they created different societies: some were hunter-gatherers, some were farmers, some possessed metal tools and weapons, some did not, etc. These geographically divided peoples also differed in their infectious diseases, first, because their local environments and biotas differed (for instance, America was not suitable for the evolving of yellow fever, while Africa was); secondly, because the behaviors of widely separated peoples differed, so that, for instance, the aborigines of Australia did not have cities and migrated often, and therefore rats and plague were not problems in Australia until introduced via outsiders.

Extraterrestrial colonists of future generations will innocently cultivate new strains of pathogens (germs) in their remote colonies, innocently export them via freighters and ferries. At the next colonies visited the new micro-organisms will celebrate their travels with virgin soil epidemics, i.e., epidemics among people who have never experienced the infection before or within a full generation. The occasional contacts between the colonies and between the colonies and Earth will enable insular pathogens to migrate. This has happened in the 14th century when the Black Death followed the Silk Road, brand new by paleoanthropology's standards, east and west across Eurasia to China and Europe, and even to Iceland. It happened again when Columbus brought the Old and New Worlds into contact, triggering the worse demographic disaster of all human history. Space travel and colonization will alter the size, shape, strengths and functions of the bodies that we have inherited from our hunter-gatherer ancestors. The first humans on Mars, Europa, etc., will be adult technicians of one kind and another. They will yearn for recognition and promotion, not for propagation. They will be succeeded by real settlers, male and female, of similar ambitions, but who will also want to build families. These people will produce the first human babies not born on Earth. Different colonies will differ radically in environment (in radiation and gravity, for instance), thus stimulating mutation. For example, what would the pregnancy of a 60-kg woman be like on Mars, where she would weight about 20 kg due to different gravity? What would her baby be like – if it survived? My guess is that after, say, a thousand or so years of extraterrestrial propagation, distinctively Martian physical and functional differentiation will be appearing. Ten or twenty generations after that, Earthlings and Martians may be different enough to qualify as separate species. And, of course, there will be another humanoid species in the colony on Europa, another on Titan, etc. Terrestrial Homo sapiens will have cousins, a situation we have not known since the demise of the last Neanderthal. The challenge to our self-image and therefore to our ethics and behaviors will be as great as it was in the years following 1492 when Columbus and his successors had to decide whether to consider American Indians as fellow human beings or not.

[8] That was the lowest of contemporary estimates, based on eyeball estimate of the coastal population and probably ignoring the considerable numbers living in the interior. Some demographers have recently judged the total population there at the moment of contact at 800,000.

1.3 Future encounters: learning from the past?

Luca Codignola

1.3.1. Discovery, encounter, meeting, contact: old wine in new bottles

Historians may well be accused of adding a rather sombre note to a debate that takes the necessity (but not the inevitability) of space discovery and exploration for granted such as it is presented in this book. Among historians of the early Atlantic world, such as myself – let alone anthropologists, ethnologists, and political scientists – discovery and exploration have recently become unfashionable, if not altogether disreputable, subjects of study among historians. Even the term "discovery" and its apparently more correct substitute, "encounter" have fallen into disgrace, because such terms allegedly give only a European point of view. In fact, I have myself used the term "encounter" for this article simply to avoid the wrath of the scholarly community – although I still prefer the world "discovery," which is at least explicit. Indeed, the notion of "encounter," as applied to Christopher Columbus's 1492 navigation and the so-called "meeting of the two worlds," implies that "encounter" first took place between Europeans and American aboriginal peoples, as if other encounters between communities, ethnic groups, nations, peoples and cultures had not taken place prior to 1492 – a notion that is patently false.

We all recall director Steven Spielberg's 1977 moving picture, *Close Encounters of the Third Kind* (1977), in which human reaction to physical evidence of a benign alien presence on Earth is depicted. Twenty years later, another Hollywood moving picture, Robert Zemeckis's *Contact* (1997), based on astronomer Carl Sagan's novel, used the word "contact" instead. This was then – and still is to this day – the accepted buzzword that replaced both "discovery" and "exploration" and, up to a certain point, "encounter." Its proponents maintain that the idea of "contact" is more appropriate than its predecessors to make audiences and readers understand the points of view of communities, which meet in any part of the globe. This very sociological emphasis on the meeting of communities, as opposed to races, peoples, or nations, has now completely replaced the genres of national epic and even scientific reportage that had been the hallmark of discovery and exploration studies.

The origins of this shift from discovery and exploration to contact studies can be traced back to the late 1960s. The process came to a climax on the occasion of the Columbus Quincentenary in 1992, when "almost all the assumptions behind the European expansion . . . were fiercely interrogated and in many cases repudiated."[9] This new attitude towards European expansion has spread even within the famous Halkuyt Society, which was established in London in 1846 to study the "records of voyages, travels and other geographical material of the past," but that seems now principally devoted to "illustrate alleged human and environmental disasters caused by European out-thrust and cultural encounter." In the words of one of its leading members, British historian Paul E.H. Hair: "What was [in the past] too loudly trumpeted as praiseworthy is now not uncommonly seen as contemptible; what was considered positive and a global gain is widely . . . interpreted as negative and a loss for humankind."[10] From all this, there seems to be only one possible conclusion. The concepts of discovery and exploration should be applied only to places where no human being had ever set foot prior to the arrival of the discoverers and the explorers. This narrow definition leaves us with Antarctica, much of the Arctic region, previously inaccessible mountain and desert areas, and, finally, outer space.[11]

In the latter, however, as opposed to the top of the Himalaya ranges, encounter with alien entities could indeed take place – tomorrow, next year, in future generations – if it has not already taken place in the past, without myself being aware of it. I know that among scientists and specialists in the field, opinions vary as to the possibility of such encounters and especially their timeframe. However, as far as I can tell, no one seems to simply exclude such a possibility, if only because we all recall from our elementary school days the somewhat caricature-like depiction of the Spanish scientists who did not believe Columbus' claim that he could reach Asia from the other side. (That they were right and Columbus was wrong has been obliterated by the fact that America was actually discovered in the process.) Our technology might still be primitive in that regard, but whoever or whatever is out there might simply reach us, as opposed to us finding them aboard some *Enterprise* ship in *Star Trek* fashion. (Paolo Musso's article in this book, however, seems to exclude even this possibility.)

Indeed, can humankind prepare itself for such a possible encounter with alien entities in outer space by learning from the experience of past encounters on this planet, for which there is plenty of evidence, especially with regard to the past one thousand years or so? Let us review some of the main issues involved in these past encounters before getting onto a discussion on whether such a preparation could really an option for humankind. Overall, there are two main issues involved in past encounters, at least as far as the western world is concerned. The first is ideological and the second biological.

1.3.2. Contact: the ideological issue

With regard to ideology, the ideological framework for encounters was provided by Christianity and by the doctrine and practice of the Christian church – which in the late medieval and early modern ages was the entity that defined both Europe and the Western world. The reaction of the Holy See to the discovery of America falls entirely within its long-established tradition. In fact, there is little difference between the papal bull, *Orthodoxe fidei propagationem* (1486), issued just before Columbus's earliest voyage, the two bulls known as *Inter caetera* (1493), issued as a reaction to it, and another bull known as *Veritas ipsa* or *Sublimis Deus* (1537), issued about one generation from the earliest contact. In spite of the fact that a new continent, let alone an entirely new race, had been discovered – potentially millions of individuals – of which there was no trace either in the sacred texts or in the teachings of the Christian church, the human nature of the new race was never doubted.

In fact, the process of European expansion, as well as the attitude of the Holy See, had rather distant origins. This process can be dated back at least to the Portuguese conquest of the fortress of Ceuta in Morocco (1415), and the ensuing discovery of the Atlantic and Equatorial islands. Or earlier, to the 1340s, when the Canary Islands, with its local population, were discovered. Or even earlier, the announcement of the First Crusade in 1095. As is well shown by U.S. historian James M. Muldoon, a distinguished contributor to this book, the Holy See and the Christian Church did not wait for the discovery of America to address the issue of how to deal with communities that did not base their existence on the same principles. There had been a debate revolving around the ill-treatment of non-Christian peoples at the time of the Crusades. Furthermore, there had been an attempt on the part of pope Innocent IV in the 13th century to sort out the Holy See's legal precedents concerning non-Christian peoples. All in all, the discovery of America represented a minor psychological trauma for the Christian church, especially when compared to the arrival of the Mongols into the world scene in 1221. At the time, Europeans shuddered at the realization that the Earth was not inhabited mainly by Christians and Muslims, besides some minor barbarian tribes living on its fringes, but that, on the contrary, the opposite was true, showing that Christians were a tiny minority surrounded by millions of menacing pagans.

At the time of the Mongol invasion as well as at the time of the discovery of America adjustments had to be made, both at the theological and at the practical level, but the Christian church did not go through any major crisis – especially when compared to the one it went through on account of the Protestant Reformation, which was almost contemporaneous to the discovery of America. Priorities did not change. Union with the Eastern Christian churches remained the first one, later to be coupled with the reclaiming of the Protestants. The reconquest of the Muslim lands came immediately second. Third place was held by the pagan peoples, encircling the Mediterranean Sea, among which the Oriental peoples constituted the vast and most interesting priority.[12]

1.3.3. Contact: the biological issue

The second issue involved in past encounters is biological. We all are familiar now with the locution "Columbian exchange," signifying what happened, in biological terms, when Europeans met Americans in and after 1492 – the exchange of microbes, bacteria, and pathogens, to use a non-scientific terminology, which was made possible by the migration of plants, animals, and human beings first across the Bering Strait and then across the Atlantic Ocean. Later, this locution came to be used also in conjunction with other "new worlds" such as certain areas of Africa and, especially, Australia, New Zealand and the Pacific Islands. This is now such a pervasively used concept that we tend to forget that it began with a book published in 1972, titled *The Columbian Exchange*, authored by then 41-year-old U.S. historian Alfred W. Crosby, Jr.[13] The Columbian Exchange thesis was later modified, sharpened, and even challenged by later practitioners in the several disciplines it touched, but a generation after it was first formulated it remains a fundamental point of reference. Allow me to recall the main points on which there seems to be substantial agreement among scholars.[14]

In 1492 the Americas were not an empty land, but indeed, they were almost as populated as Europe.[15] Except for very sparse encounters in the Newfoundland region, most of the contact activity took place in areas where the American population was overwhelmingly superior to the number of newcomers that the Spanish ships were able to deliver to the Caribbean and Central America. Between 1492 and 1573, the period historians pompously refer to as the Conquest, the Spanish, who did not arrive in great numbers, wiped out some American nations (for example in the Caribbean islands), subjugated a number of them, and took control of some of their useful territory. But a conquest it was definitely not, because the Spaniards never had the manpower, the willingness to complete it, or the adequate resources to do so.[16]

The issue of the initial clash, however, remains. How did so few Spaniards defeat such a variety of American societies, which inhabited such extensive territories and were often densely populated, within comparatively little time and with so few men? Over the centuries, historians have offered a number of seemingly plausible explanations, none of them fully satisfactory. There is the military explanation:

the Spanish were vastly superior because they used firearms, horses, and dogs; they were all trained soldiers; and used organized terror as a matter of course. There is the psychological and mythical explanation: local cultures mistook the Spanish for deities and were awestruck by their appearance and power, so that they did not immediately react to the deadly threat they posed. There is the political explanation: a number of American nations tried to use the Spanish newcomers as military allies against their enemies or the ruling elites. All these reasons, or combinations thereof, have some validity with reference to specific instances. They fail, however, to take into account the overall length of the initial military conquest, which, rapid as it was, lasted for almost three generations. They also belittle the Americans' ability to learn, adapt, and react. Conversely, they overestimate the effectiveness of the Spaniards' firearms, the long-term psychological impression made by their appearance, and the political unity of the *conquistadores*.

Furthermore, why is it that nothing of that sort happened in Africa, where the Portuguese and the Spanish might have profited from the same advantages? In fact, the real difference between contact in Africa and in the Americas was that in Africa Europeans proved unable to survive in the new biological environment (in which adult Africans already were lucky survivors in a catastrophic biological environment that killed off most of each new generation), whereas in the Americas it was the Americans who proved unable to survive[17] (Curtin, Davies). The latter followed the path of the natives of the Canary Islands one century earlier, who were shattered by the deadly violence of the encounter with the pathogenic microbes that the Europeans had, unwillingly and unconsciously, brought with them. After thousands of years of separation from the rest of the world, the Americans had lost all the biological defences, which allowed Europeans, Asians, and Africans to resist simple illnesses such as cold and influenza with some possibility of success. For the same reason the Americans were sitting targets for infectious diseases which probably were smallpox, mumps, whooping cough, chicken pox, measles, and scarlet fever.

From a biological point of view, the consequences of contact were as rapid as they were catastrophic. In the Caribbean and Central America in general (smallpox first appeared in Santo Domingo in 1518); the population diminished by more than 900 per thousand in only the first quarter of the century following contact. It has been calculated that the Mexican population collapsed from 25,200,000 in 1518 to 750,000, its lowest figure, in 1622. The disappearance of the Americans was so appallingly rapid and widespread that every European observer noted it. People were dying inexplicably, in unimaginable numbers and in altogether new ways. It was as if the balance between animals, plants, human and spiritual being which until then had underlain the physical and intellectual universe of the Americans had been broken. The Americans who survived the Spanish microbe invasion began to lose any will to live. In short, this biological imbalance achieved negative demographic results amongst the Americans that no Spanish violence, no matter how conscious, ruthless and well organized could have ever achieved.

Two correctives need to be added immediately, however, to a "catastrophe" reading, which might quickly transform into a quasi-biological determinism. The latter would not explain, for example, why some American nations survived the early impact of contact and others did not. The first corrective is that Central American epidemics did not spread, swiftly and automatically, throughout the rest of the Americas. For example, in North America the first epidemic we know of broke out in 1616 on the coast of present-day Maine, and the first outbreak of smallpox was as late as 1633. The second corrective is linked to the fact that, contrary to the romantic image of a pre-contact American world without significant illnesses, Americans were short-lived and suffered from malnutrition, anemia, arthritis, osteoporosis, blindness, cavities, skin diseases, tubercular infections, pneumonia, and treponematosis.[18] This means that their communities reacted differently to the impact with European germs, depending on their state of health. Some were destroyed, such as the Caribs and the Arawaks. Others overcame the initial trauma relatively easily. Although specific explanations are harder to come by, we can date the initial demographic recovery of the Americans as early as the mid-17th century.

Once we have taken into account these two main issues involved in past encounters – ideology and biology – which we have described mainly with regard to the Americas, but that could very well be applied to a number of other experiences, the history of the encounter between Western civilization (or Christianity, or Europe, as we have defined it so far), is mostly a long history of reciprocal adjustments. Over the long run, it is rather easy to depict this encounter as a progressive conquest in which Europeans got the upper hand and the so-called non-Europeans were either wiped out, enslaved, or exploited. In the case of Asia and Africa, this is simply untrue, as proven quite simply by demographic data. In the case of the Americas and the South Pacific, this unfortunate process did indeed take place, but when carefully examined over short-time spans or geographically-limited regions, this simplistic explanation belittles the non-European peoples' ability to react, adjust, create, and profit from the European presence.

1.3.4. Has history prepared us for contact?

Let us now go back to the future and briefly discuss whether past experiences of humankind can teach us how to be better prepared to meet the challenges of future encounters in outer space – or future encounters with entities visiting us from outer space, whatever the case may be. Based on past experience, on paper, two main options await us. The first is biological. When we meet new entities, one of the two does not survive the challenge – or at best comes out of the initial encounter very weak, physically disabled and psychologically shattered. We hope humankind is not going to be the loser in this initial conflict, in which voluntary decisions and individual agency is not at play. If we come out the winner, however – as indeed we all hope – our biological superiority might carry with it some feeling of culpability that we may later overcome by extolling the virtues of the "others". Once, of course, they have been exterminated or reduced to a non-challenging role by the microbes of our ancestors. It has, indeed, happened in the past.

The second option is ideological. If, for any reason, biology does not represent a major factor in this encounter, then the range of options opens up to a variety of possible experiences, from the good and wise little guys of *Close Encounters of the Third Kind* to the nasty and destructive killing machines of *War of the Worlds* (2005) – an almost 30-year itinerary that perhaps is more representative of Spielberg's growing pessimism rather than of a more educated reflection on future encounters. Most likely, both sides will experience variations in the behavioural categories of adjustment, adaptation, and compromise, all of them bringing about change on both sides, or, at least, on our side. But to what extent would ethics be involved in the process? Would the other side be visibly superior to us from a moral point of view, so that we would be placed in a real quandary, having to decide whether to keep our identity as a race or to lose it by adhering to that of the newcomers? But then, what ethics are we really talking about? Our or theirs? Allow me a reference to yet another moving picture, Kevin Costner's *Dance with Wolves* (1990). Was Union Army Lt. John Dunbar betraying his "race," or was he actually fulfilling its moral commandments by siding with an alien "race," the Sioux, that practiced them at their fullest extent? When we bring in outer space, is there a chance that we share the same moral commandments by coming from different sides of the universe?

But if one really wants to know what the study of history has taught us about preparing for future encounters with entities from outer space, the truth of the matter is that there is no way humankind can be ready for them – if, and when, these encounters take place. No weapons can be readied, because we do not know our enemies. No God can be brought in, because we would not know whether it is ours or theirs. No production can be implemented, because we do not know what can be useful. No exchange can be envisaged, because we do not know what is needed. Whenever the time of encounters came in the past of humankind, neither side was ready. The antibodies of the Arawaks of the Caribbean were not ready to identify and neutralize the foreign bacteria and viruses brought in by the Spanish.

The Spanish were not ready to conquer and manage such a vast land in so little time. The Algonquins were not ready to trade their beaver skins that they had used and consequently greased throughout their winters for valuable French glass beads and metal pots. The French were not ready to bring their trinkets to the New World and bring back valuable, shiny furs. The Irish were not ready to eat potatoes nor were Neapolitans ready to use tomatoes for their pizzas. Finally, the Chinese were not ready to kill themselves by the millions by smoking filthy cigarettes produced with Indian tobacco. Yet humankind, in the past, has proved to be able to survive, through sometimes at great costs, through adjustment, adaptation, and compromise. We trust that will happen again, when the time comes, although at what cost, no one is able to foresee.

---

[9] Warkentin, Germaine, ed. 1995. Critical Issues in Editing Exploration Text: Papers Given at the Twenty-eighth Annual Conference on Editorial Problems: University of Toronto 6–7 November 1992. Toronto, Buffalo, and London: University of Toronto Press.

[10] Bridges, Roy C. and Hair, Paul E.H. eds. "The Hakluyt Society and World History." In: Compassing the Vaste Globe of the Earth: Studies in the History of the Hakluyt Society 1846–1996: With a Complete List of the Society's Publications. London: Hakluyt Society. 1996. (Reston: American Society of Civil Engineers).

[11] Codignola, Luca. "North American Discovery and Exploration Historiography, 1993–2001: From Old Fashioned Anniversaries to the Tall Order of Global History?" Acadiensis 31 (2002): 185–206.

[12] Codignola, Luca. "The Holy See and the Conversion of the Indians in French and British North America, 1486–1760". America in European Consciousness, 1493–1750. ed. Karen Ordahl Kupperman. Chapel Hill: University of North Carolina Press, 1995. pp. 195–242.

[13] The editors are grateful to Dr. Alfred W. Crosby for accepting to deliver the closing key-note address at the seminar that took place in Genoa, Italy, on 22–23 March 2007 ("Humans in Space. A Humanities Assessment of the Implications of Space Sounding and Exploration"), which started off the process that eventually led to the present book.

[14] Crosby, Alfred W. The Columbian Exchange. Biological and Cultural Consequences of 1492. Westport: Grenwood Press, 1972. Crosby, Alfred W. Ecological Imperialism. The Biological Expansion of Europe 900–1900. Cambridge: Cambridge University Press, 1986. Crosby, Alfred W. Germs, Seeds, and Animals. Studies in Ecological History. Armonk, NY.: M.E. Sharpe, 1994. Cook, Noble D. Born to Die: Disease and New World Conquest, 1492–1650. Cambridge: Cambridge University Press, 1998.

[15] Denevan, William M, ed. The Native Population of the Americas in 1492. Madison and London: University of Wisconsin Press, 1992. First Published in 1976.

[16] Kamen, Henry A.F. Spain's Road to Empire. The Making of a World Power, 1492–1763. London: Allen Lane, 2002. Published in the United States in 2003 as Empire. How Spain Became a World Power, 1492–1763. New York: HarperCollins.

[17] Curtin, Philip D. Death by Migration. Europe's Encounter with the Tropical World in the Nineteenth Century. Cambridge: Cambridge University Press, 1989. Davies, Kenneth G. "The Living and the Dead: White Mortality in West Africa, 1684–1732". Race and Slavery in the Western Hemisphere. Quantitative Studies. Eds. Stanley L. Engerman and Eugene D. Genovese. Princeton: Princeton University Press, 1975. 83–98.

[18] Aufderheide, Arthur C. "Summary of Disease before and after Contact:" Disease and Demography in the Americas. Verano, John W. and Ubelaker, Douglas H. eds. Washington, D.C.: Smithsonian Institution Press. 1992. pp. 165–166.

1.4 Are we alone? Searching for life in the universe and its creation

Gerhard Haerendel

1.4.1. Introduction

This contribution is based on a presentation given at the workshop in Genoa, preceding the conference "Humans in Outer Space". This presentation ended with six conclusions. They will now appear as six theses and at the same time as a concise outline of this paper. The subsequent elaborations will be inadequately short in view of the vastness of the subject, the uncertainties pervading almost all conclusions, and their enormous philosophical consequences. Moreover, a large portion of subjectivity will adhere to the conclusions and little room will be devoted to existing controversies.

Six theses:
1. Life is common in the universe.
2. Life elsewhere in the solar system must be microbial.
3. Research on extrasolar planets may eventually reveal the existence of life-supporting environments.
4. We are physically confined to the solar system.
5. SETI is the only way to find out whether we are not alone in the universe.
6. Sending messages into the galaxy ("active SETI") may be regarded as a moral obligation.

1.4.2. The ubiquity of life

The conviction that our planet is not unique in the universe with respect to the existence of life is founded on two arguments, the almost "immediate" appearance of life on Earth, after it had cooled down, and the universal availability of carbon, hydrogen and other light elements, out of which all living organisms on Earth are composed. This argument pertains to the existence of simple mono-cellular forms of life. Higher, multi-cellular forms may also be common in the universe, but in view of the nearly 3 billion years it took for them to develop on Earth, one is

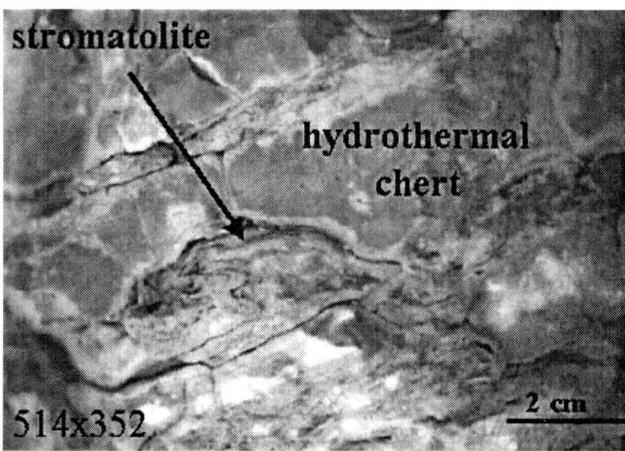

Fig. 1. *The first distinct traces of microbial life, mineralized bio mats or stromatolites created by photosynthesising bacteria, are 3.5 billion years old (source: Westall, F. et al. The 3.466 Ga Kitty's Gap Chert, an Early Archaean microbial ecosystem. In: Processes on the Early Earth (W.U. Reimold and R. Gibson, Eds.), Geol. Soc. Amer. Spec Pub. 405 (2006). pp. 105–131).*

inclined to postulate the need of long periods of special, stable conditions for their appearance.

The solar system was born 4.6 billion years ago out of a molecular cloud. During the initial 700 million years, the interior structure of our planet formed by differentiation, while the surface was essentially a magma ocean, heated from the inside by the decay of radioactive elements and from the outside by an enormous impact rate of small bodies, which were richly abundant in the early solar system. The oldest known rocks formed 4.4 billion years ago. The first distinct traces of microbial life, mineralized bio mats or stromatolites created by photosynthesising bacteria, are 3.5 billion years old (Figure 1). However, the isotopic composition – ^{12}C versus ^{13}C – of carbon globules in metamorphic rocks of 3.8 billion years ago, which were discovered in south-western Greenland, is strongly indicative of enrichment through life organisms.[19] This finding and its not uncontroversial interpretation as an indicator of life, forms the basis of our statement of the "almost immediate" appearance of life on Earth.

Carbon is abundant in the sun, planets, dust, other stars and molecular clouds. Carbon can form stable chain and ring molecules, which are the components of the more than 10 million organic molecules known. By comparison, only about 100,000 inorganic molecules are known. Apart from carbon, no other element has an equivalent molecule-forming ability. This is the essential argument for the prevailing conviction that life elsewhere must be based on carbon–hydrogen compounds and is therefore possible anywhere in the universe where the right conditions of habitability prevail.

While it seems likely that simple forms of life will spontaneously appear wherever the conditions of habitability are met, the time needed for the development of multi-cellular life must be much longer. On Earth, the first may have needed only about 100 million years, while the latter took 3 billion years. Habitability[20] includes the existence of a sufficiently dense atmosphere, water, and average temperatures within a moderate range that keep water liquid most of the time. In addition, the planet must enjoy orbital stability and not too extreme seasons or ages. During the 3 billion years of evolution, the existing mono-cellular life severely modified the environment, the most important change being the enrichment of atmospheric oxygen and ozone and the depletion of carbon dioxide and methane. In the first place, it involved photosynthesis, and subsequently the discovery of the advantages of oxidation versus fermentation of food, of symbiosis, of formation of colonies, of sexuality and meiosis. Of course, under other initial environmental conditions life might have taken a different course, but with the (incomplete) knowledge of only one example, our Earth, we are compelled to postulate the need for a period comparable to the lifetime of a planet before the symbiosis of mono-cellular (bacterial) life evolves into complex multi-cellular organisms. From here to the appearance of mobility, five senses, brains and intelligence, it may be a comparatively short step. On Earth, it took only one-seventh of the total time of the existence of life. All of this is naturally speculation, not without reason, but it serves our later estimates of the probability of the existence of life elsewhere.

1.4.3. Searches for life in the solar system

Mars is the most promising planetary body in the solar system for housing extinct or even, although much less likely, extent life. The Jovian satellite, Europa, with its subsurface ocean is another candidate. Mars is located, perhaps marginally, in the zone of habitability of the solar system. Strong evidence exists that Mars once possessed a dense atmosphere and abundant water. This evidence comes from the finding that strong erosion features, such as dried-up river channels and dendritic valley networks are found on craters more than about 3.5 billion years ago. Why Mars lost that atmosphere is still not understood and certainly one of the most intriguing questions about the evolution of our solar system. How much water is present today in the form of subsurface ice is not well known. It is expected to be much less than during that initial period. The appearance of new local erosion features during the time Mars has been under close surveillance is evidence for episodic water flows, probably caused by local heating events.

After the discouraging outcome of the visit of Mars by the Viking landers in 1976, it was the discovery of potential traces of mineralized microfossils on the Mars meteorite ALH84001 in 1996 what revived the interest in in situ research on the planet. An armada of orbiters and rovers have since then been sent to Mars, and a wealth of new knowledge of the surface and interior conditions has been created. However, as yet no compelling evidence for traces of life has been found. To a large extent this may be a matter of the analysis methods employed and of their sensitivity. Although Mars presents itself with a rather hostile environment, there is nothing known that would preclude the appearance of life, at least within the first few hundred million years. However, our arguments above about the long time needed for the evolution of more complex forms of life strongly suggest that, if at all existent, life on Mars must have been microbial.

Great expectations are now placed on the ExoMars mission of the European Space Agency, which is scheduled to be launched in 2013. The key element will be the Pasteur rover, which will carry a set of very sophisticated instruments to search for signatures of life, both at the chemical and structural level. The Pasteur exobiology package will be complemented by a set of geophysical and environmental experiments.

The perhaps most important analytical instrument is the Urey experiment, composed of two parts, the Mars Organics Detector and the Mars Oxidant Instrument.[21] The first will extract, purify and then analyze by electrophoresis organic compounds, such as amino acids and nucleobases, with a sensitivity exceeding that of the Viking instruments by a factor of 10,000. Figure 2 shows an

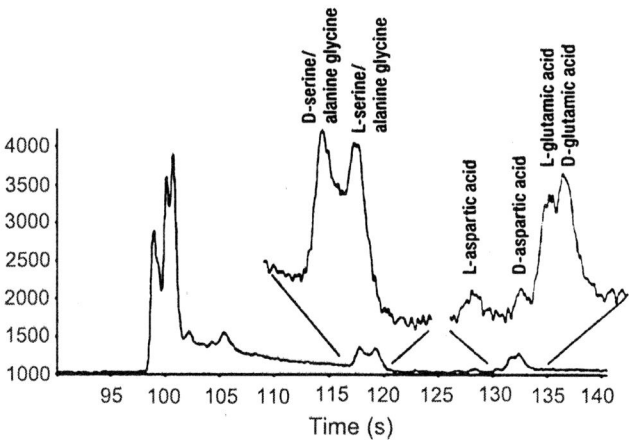

Fig. 2. *Composition and chirality of amino acids by the capillary electrophoresis unit of the Urey instrument (source: Bada, Jeffrey L. et al. Mars Organic and Oxidant Detector Searching for Signs of Life on Mars ESA Pasteur/ExoMars Mission. Solar System Exploration. 2007. http://astrobiology.berkeley.edu/projects.htm).*

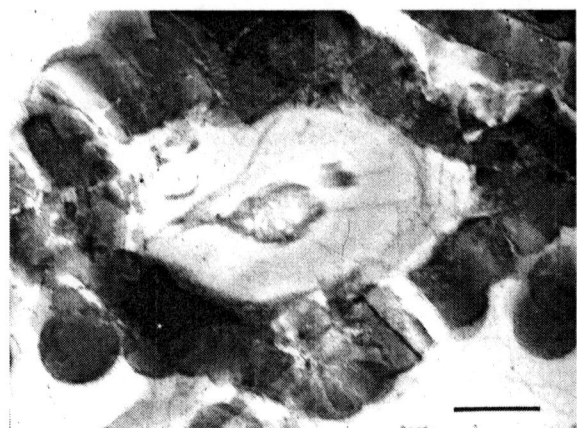

Fig. 3. *Outer cell envelope of an experimentally mineralized bacterium. Degraded organic matter is trapped in the mineral matrix. The bar is 0.5 mm long (source: Westall, F., Boni, L., and Guerzoni, M.E. "The experimental silicification of microbes". Palaeontology 38 (1995): 495–528).*

analysis example of the composition and chirality of amino acids by the capillary electrophoresis unit of the Urey instrument. The Mars Oxidant Instrument will be used to determine the oxidative characteristics of the samples in order to understand the role of oxidation reactions on the survival of organic compounds in the Martian regolith. It is thought that such reactions were responsible for the lack of any evidence for the existence of life organisms in the Viking experiments.

Of the various physical instruments for structural analysis on the macroscopic and microscopic level, I would like to mention the Microscope,[22] which will be able to detect the morphology of mineralized microbial material at grain-size level. Figure 3 gives an idea of the degradation of the original bacterial structure in the mineralization process and of the detection problems. Spectroscopic devices will assist the analysis of the involved degraded organic macromolecules.

1.4.4. Extrasolar planets

More than 250 extrasolar planets have been identified thus far, and all are located within a range of 800 million light years. Almost all of them are gaseous and of Jupiter size. Life-bearing planets must have solid surfaces, stable orbits and reside in the zone of habitability.[23] None of the planets discovered up to now fulfil all of these conditions. Their orbits are overwhelmingly eccentric, implying extreme climatic variations. The detection methods available to date favour high masses and orbits close to the central star. Therefore, it is no surprise that no candidates for housing life have yet been identified. Figure 4 contains masses and distances of the

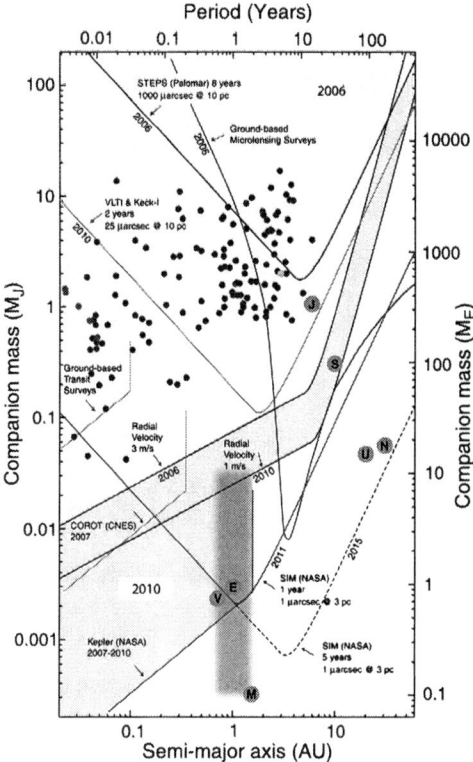

Fig. 4. *Exoplates and detection matters (source: http://en.wikipedia.org/wiki/Extrasolar_planet).*

known exoplanets and also the limitations of the various detection methods.[24] Among the four indirect methods there is none capable of discovering Earth-like planets within the zone of habitability, i.e. at distances of roughly one astronomical unit from the centre of a sun-like star.

The art of finding extrasolar planets is still in its infancy. The enormous interest in the question of whether life exists elsewhere in the universe will drive the development of techniques that will eventually enable not only the discovery of Earth-like planets at habitable distances from the star, but also allow a rough spectroscopic analysis of their atmospheric composition. However, this requires great technological advancement and huge expenditures. One promising technique is the "nulling interferometry". Figure 5 shows an artist's impression of the set up. The IR light received by several satellites will be collected on the focal plane of a central spacecraft and made to interfere (by phase shifts) in a destructive manner, so that the light of the central star is blocked out and that of the small planet in its neighbourhood, at one-millionth of the intensity, can be separated out. When the

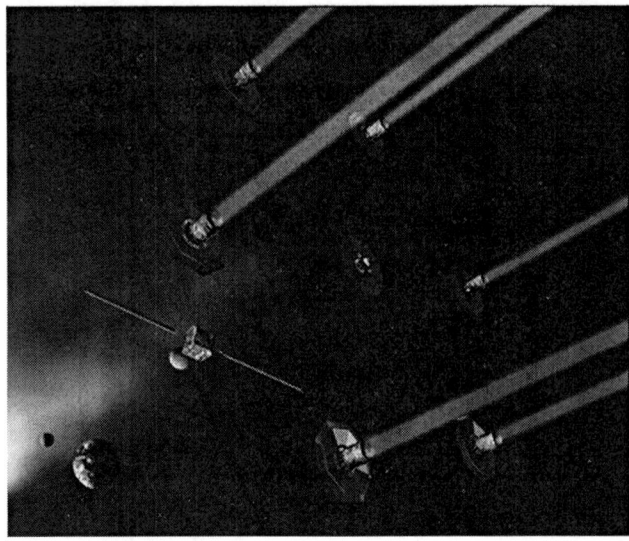

Fig. 5. *Artist's impression of 'nulling interferometry' (source: Darwin, looking for Earth-like planets. ESA Science and Technology. http://darwin.esa.int/science-e/www/object/index.cfm?fobjectid=32587, 2008.).*

sensitivity of this yet unproven method will have become sufficiently high, one may succeed in finding the spectral signatures of key atmospheric constituents such as ozone and water. Figure 6 shows a comparison of the IR spectra of Venus, Earth, and Mars.[25] Besides CO_2, which is present in all three atmospheres, only that of Earth contains the absorption bands of ozone and water vapour.

With all the new insights to be eventually gained with the availability of such advanced detection methods, one can hardly hope to discover direct signatures of life. Only evidence for habitability and indirect signatures of a life-supporting environment may be found. However this may turn out to be invaluable for guiding searches for life by Search for Extra-Terrestrial Intelligence (SETI) activities, as will be discussed below.

1.4.5. Confinement to the solar system

Science fiction has spoiled the conception of the public about the borderlines of the physically possible and the impossible. It is a sobering exercise to calculate the energy demands of a space mission covering stellar distances within the lifetime of a human being. As an example, we chose a one-way trip to our nearest neighbour Alpha Centaury, with a travel time of 40 years. In order to cover the distance of 4 light years within this time, one needs to travel at an average speed of 10% of the

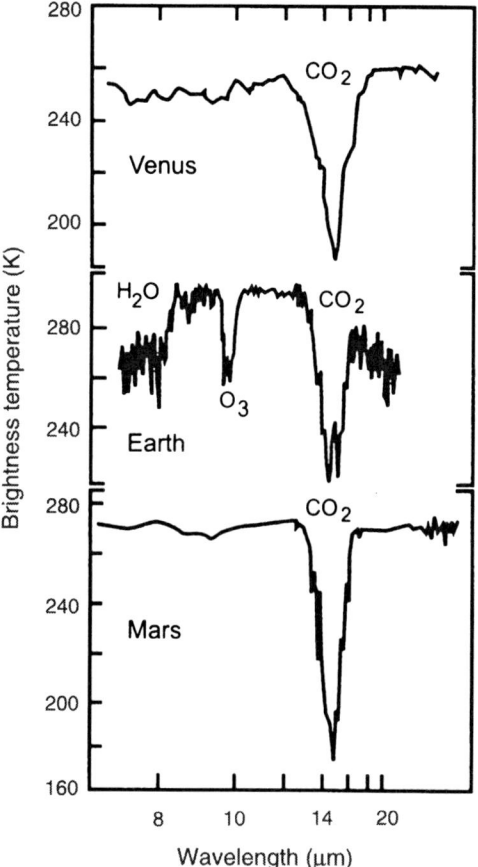

Fig. 6. *Comparison of the IR spectra of Venus, Earth, and Mars (source: Angel, Roger P. and Woolf, Neville. J. "Searching for Life in Other Planets". Scientific American Apr. 1996: 46–52).*

speed of light. The special theory of relativity tells us that to accelerate a space ship of, say, 1000 tons, one needs energy of 4.5×10^{20} J. This is equivalent to the total present annual energy consumption of the world. If one wanted to supply this amount of energy in the most concentrated way presently available, namely as nuclear fission fuel, one would need 7000 tons of it.[26] Adding the mass of the nuclear reactor, of the engine and everything else, one would have to apply a large factor for the total mass. Although, in principle, the laws of nature do not rule out such an endeavour, it is practically impossible.

This short mental exercise gives us a sad insight, namely that humankind will forever be confined to the solar system. Although such speculations exist, it is hardly conceivable that other civilizations, if there are any, would have much more powerful resources at their disposal and have such a long life expectancy than human beings as to enable them to cross stellar distances at significantly lower speeds with the corresponding much lower energy requirements. As a consequence, one must conclude that highly developed complex living organisms must be lonely, although not necessarily alone, in the universe, trapped in their respective solar systems. The same may not hold, though, at the molecular level. Organic molecules, which are richly abundant for instance in molecular clouds, may live very long and travel large distances on tiny dust grains. Whether one can derive from that a general panspermia theory for the origin of life remains, however, questionable. It is understandable that the growing realization of the confinement of humankind to our solar system nourishes the wish to explore and conquer the accessible bodies to the maximum feasible and reasonable extent.

1.4.6. Communication with other civilizations

One of the greatest goals of solar system research is the search for life. We may actually succeed and one day find forms of life similar to ours or even very different ones, on Mars or other bodies in the solar system. This would undoubtedly greatly widen our insights into the nature and origin of life in general. However, we might forever be without concrete knowledge of life outside our solar system, unless we succeeded in communicating with other civilizations. The subject of interstellar communication consists of three critical elements, the existence or non-existence of other civilizations, the possible range of communication with electromagnetic waves, and the willingness to communicate. The first problem can only be solved once we have received an intelligent signal from another celestial body. The second element, the communication technology, is already well advanced on Earth. We can already now send messages over distances in the order of 1000 light years that other civilizations, on equal or more advanced level of development, could easily receive. A significant extension of this scope for communication may be realistically expected as our technologies advance. The main problem is the last point, the willingness to truly engage oneself in an attempt to send messages. And this problem does not only exist for our civilization.

Let us first look into the second issue, which is entirely technical. The most promising way of sending interstellar messages is by radio waves in the spectral bandwidth between 2 and 30 cm wavelength. This radio window is limited, on one side, by galactic noise and, on the other, by atmospheric absorption. It is the band in which most of the SETI activities are undertaken. Inside this band lies the famous 21 cm hydrogen line, which offers one of the most powerful tools for investigating the distribution of interstellar gas in the galaxy. The astronomical prominence of this line and the low absorption within the surrounding radio band will also be noticed by other civilizations. Like us they are likely to choose neighbouring frequencies when sending interstellar messages. An impressive amount of SETI activities have been undertaken since Project Ozma in 1958 using various radio antennas and strategies, also involving radio amateurs all over the world in analyzing the signals received. However, no effort has yet been made to continuously and systematically search the whole sky. Only certain promising stellar objects have been selected for pointing the telescopes. Progress in identifying Earth-like exoplanets will eventually provide a powerful means of guiding the search.

In order to get an idea of the effort needed for a full approach, let us look at the example of a radio dish of 200 m diameter with a transmitted power of 1 MW. If one chooses a narrow bandwidth of the order of 1 Hz, the signal-to-noise ratio at a distance of about 2000 light years will be sufficient for its detection by an equally large telescope. There are, however, two problems. The motion of the Earth, its rotation as well as its revolution around the sun, will Doppler shift the received frequency from any fixed direction. To cope with this phenomenon radio receivers with millions of narrow-band channels have been developed which permit one to continuously tune the received frequency so as to counteract the Doppler shift. The other problem arises from the narrow angular width of a beam at several centimeter wavelength transmitted by a 200-m telescope. It has the size of only 6% of the lunar disk. Two million pointings would be necessary to cover the whole accessible sky. Other civilizations would have the same problem when sending messages. They may not point towards Earth every day. Thus, there is an additional timing problem. This exercise shows clearly the size of the complete SETI effort. Feasibility in principle does not lead to immediate realization in spite of the enormous philosophical relevance of the issue.

The main hindrance to interstellar communication arises from its low probability of success. This is best demonstrated by an evaluation of the Drake equation, which is a simple estimate of the probability of N Earth-like planets existing in our galaxy and housing a civilization capable of sending interstellar messages.[27] Actually, N is composed of a product of various probabilities defined below. For simplicity, I have attached my personal evaluations to the definitions of these probabilities.

$$N = N_s/L_s \times f_p \times n_e \times f_l \times f_i \times f_c \times L$$

where N_s is the no of sun-like stars in the galaxy $= 4 \times 10^9$; L_s is lifetime of a sun-like star $= 10^{10}$ years; f_p is fraction of stars with planets $= 1$; n_e is no of (habitable) Earth-like planets $= 0.01$; f_l is fraction with life developing $= 1$; f_i is fraction with intelligence developing $= 0.5$–0.01; f_c is fraction with communicating ability $= 1$; L is the lifetime of such a civilization $= 10^3$–10^5 years.

The concrete numerical values of my personal estimates are certainly debatable. They are only meant to underline the extent of our lack of knowledge. Other authors have arrived at more optimistic or more pessimistic estimates.[28] The first three numbers may not be far from reality, if one takes an appropriate definition of "sun-like". Since we have no observations yet of Earth-like exoplanets, my estimate for n_e is rather cautious. Further progress in astronomical research will yield somewhat more secure estimates. n_e may even increase. It has been argued above that the appearance of life is highly likely for a planet located in the habitable zone. Hence, I have used a probability of 1, i.e. 100%, which may be a bit optimistic. However, the development of complex forms of life may take a large fraction of the lifetime of the planet, but if they exist, the step towards the appearance of intelligence may be short, only a few 100 million years. The combined probability is very difficult to assess. Having no other information, we may consider the path life has taken on Earth either as typical or as unlikely. The range of numbers given above is meant to cover these two extremes. The greatest uncertainty, however, is related to the lifetime of a civilization with interstellar communication ability. Our only known example has existed for only about 50 years. While the greatest risks to the long-term existence of our society probably come from inside, there are also external dangers, like mass extinctions. Most likely they will not wipe out life completely, but for some time interrupt the command of high technology and resources for interstellar communication. Even the lower limit for the lifetime, L, given above, may be judged by some as being too optimistic. The upper limit certainly is.

The outcome of the above estimates is a number, N, ranging between 200 and 0.04 of such civilizations in our galaxy. The latter number means that their existence is highly unlikely, but not excluded, while the first number gives hope that within the present range of communication ability we may find one or two such civilizations overlapping with our existence. This is clearly not very encouraging.

Our last topic is a mainly cultural and perhaps a moral issue. Does it make sense to make the investments in a proper installation for passive as well as active SETI, when even in the case of a successful detection of intelligent signals, communication would involve almost forbiddingly long time spans, like many hundreds, if not even a few thousands of years? Nonetheless, it is logical not to expect to receive any message from another civilization, if we do not send messages ourselves, because what hinders us is likely to hinder others as well. These may be economical or protective considerations. Indeed, many authors have expressed fear of interstellar attacks if knowledge of our existence were to spread in our galaxy.[29] However, the considerations in the preceding section should suffice to completely exclude the existence of such dangers. The answer to the famous question of Enrico Fermi "Where are they?" is simply that wherever they are, they cannot physically get here.

Arguments against investments in a full-fledged SETI activity would then mainly be of economical nature, the costs being regarded as too high in view of the low likelihood of success. On the other hand, the enormous philosophical relevance of the question of whether we are or are not alone in the universe is a strong argument. In the long run we cannot be satisfied by just evaluating the Drake equation over the long term. Therefore, I regard it as a moral obligation of humankind to work out a good cost-efficient strategy for continuously searching for signals and sending messages, for the realization and maintenance of such activities, which are to be adapted to the progress in science and technology for the next millenniums. The costs are likely to be much lower than those for human exploration of the solar system.

1.4.7. Summary

This article began with six theses about life in the universe. Starting out from insights deduced from the history of life on Earth, I concluded that life must be universal and readily appear wherever the conditions of habitability are met. Although life elsewhere in our solar system can be only microbial, if at all existent, complex forms of life and even intelligent communities may exist within our present and, even more so, in our future range of communication. This communication can, however, only involve the exchange of information, not physical contacts. From these arguments, which are admittedly based on my personal evaluation of present days knowledge and highly laden with uncertainties, I have derived two further conclusions: Realizing that humankind will forever be confined to the solar system provides perhaps the best justification for its desire to visit and explore to the maximum feasible extent the bodies accessible therein. Even remote chances of ever receiving intelligent signals from other civilizations within our galaxy justify the necessary investments and running costs of a long-lasting effort of systematically searching for signals and transmitting messages, simply because of the outstanding philosophical and religious importance of the question of whether we are alone or not alone in the universe.

---

[19] Ulmschneider, Peter. Intelligent Life in the Universe. From Common Origins to the Future of Humanity: Principles and Requirements Behind its Emergence. New York: Springer, 2003.
[20] Ibid.
[21] Mars Organic and Oxidant Detector Searching for Signs of Life on Mars. ESA Pasteur/ExoMars Mission. Eds. Jeffrey L. Bada, et al. 2007. Solar System Exploration, University of California, Berkeley. NASA http://astrobiology.berkeley.edu/projects.htm.

[22] Vago, J. et al. "ExoMars: Searching for Life on the Red Planet". ESA Bulletin 126 (2006): 16–23.

[23] Ulmschneider, Peter. Intelligent Life in the Universe. From Common Origins to the Future of Humanity: Principles and Requirements Behind its Emergence. New York: Springer, 2003.

[24] "Extrasolar Planet". Wikipedia. 2007 http://en.wikipedia.org/wiki/Extrasolar_planet.

[25] Angel, Roger P. and Woolf, Neville J. "Searching for Life in Other Planets". Scientific American Apr. 1996: 46–52.

[26] Von Hoerner, Sebastian. Sind wir allein? SETI und das Leben im All. München: Beck, 2003.

[27] Ulmschneider, Peter. Intelligent Life in the Universe. From Common Origins to the Future of Humanity: Principles and Requirements Behind its Emergence. New York: Springer, 2003. Von Hoerner, Sebastian. Sind wir allein? SETI und das Leben im All. München: Beck, 2003.

[28] Ulmschneider, Peter. Intelligent Life in the Universe. From Common Origins to the Future of Humanity: Principles and Requirements Behind its Emergence. New York: Springer, 2003.

[29] "Shall We Shout Into the Cosmos"? David Brin's Official Website. 2006 http://www.davidbrin.com/setisearch.html#radar.

1.5 What's the story, mother? Some thoughts on Science Fiction Film and Space Travel

Thomas Ballhausen

> *"take all your reasons and take them away*
> *to the middle of nowhere, and on your way home*
> *throw from your window your record collection*
> *they all run together and never make sense*
> *but that's how we like it, and that's all we want*
> *something to cry for, and something to hunt"*
> The National: Looking For Astronauts

> *"and I'm floating in a most peculiar way*
> *and the stars look very different today* [...]
> *and I think my spaceship knows which way to go*
> *tell me wife I love her very much she knows"*
> David Bowie: Space Oddity

Within our archives preserve not only evidence of the past, but also materialized imaginations of the future are preserved there. When dealing with film as a medium, the complex, multi-layered relations between imagined, and therefore also projected times, – past, present, future – can be traced. The genres of the fantastic film, in particular the genre of Science Fiction, reflects the present while dealing with the future. So-called deviant genres and subgenres – think SF, Horror or even Porn – have always portrayed social fears and hopes and have also reflected and influenced real technical development. Therefore, Science Fiction film can also be understood as a rethinking of human space travel, opening something like a metaphorical *double-space* between real and framed space.

1.5.1. Tales about the future

Film as a medium and cinema as a structure occur as a time-machine, working in two ways: showing us the past, as well as the possible future. Here, the wonder of

35

cinematography serves as a producer of images, as an option for the concrete development of ideas – and also as a mass-appealing agent for the propagation of the ideas mentioned above. Film as the visual key medium of the 20th century is distinctly affected by a desire to innovate and an enthusiasm for (further) technical advancement. Above all, the genre of the fantastic film proves to not only have a great potential for entertainment, but also to be a visionary "impulse generator" for factual and notional technical and social developments. These developments are portrayed alternatively as positive and affirmative or as dystopic and cautionary. Thus, many classical issues of modernity can be detected in fantastic film, which, after all, did not diversify further into corresponding subgenres before the 1930s and 1940s of the last century. Among the issues discussed are questions of science, research and technology, as well as those of urbanity, artificiality and the gain or loss of control.

The idea of travelling as it has presented itself to us in literature and philosophy since the late enlightenment, can provide conceptual parenthesis here. The repercussions of the explorative-expansionary phase of early modern history are still clearly perceptible. Its visually powerful feedback has taken hold of European worlds of imagination. Little does it surprise that these imaginations often grow into nightmarish phantasmata. Consequently, topical overlapping of the exotic, the erotic and the strangely disconcerting in philosophical and fantastic travelogues is apparent even in films produced much later. However, in subsequent cinematic (ad)ventures the search for alien life is superseded by the search for *intelligent* extraterrestrial life (here we benevolently assume that the third *rock* from the sun *can* be considered as being inhabited by intelligent life-forms). At this point, another aspect of the genre becomes evident: especially films reporting of a fixed point in time that we have, in fact, already lived through, allow for a kind of critical comparison, which not only manifests itself in the utopian ideas of a particular filmic example, but also provides information about the date of origin of that example. From this angle, film once more proves to be a multi-dimensional source for a variety of historical and social questions. A look into the archive can therefore open our eyes to current trends and ongoing changes; it is a *potential* future which is shelved in our depots, and which demands a discriminative gaze – perhaps even an anamorphic perspective – in order to become visible to us.

1.5.2. Historical developments

Film, in particular the genre of the fantastic film, has to be understood as a lively texture consisting of imagination, critique and entertainment. Nevertheless, the moment of crisis, which has proven to be highly influencal on film history and the different genres, must not be underestimated. Every shock produced a bunch of films linked to the specific event – for Science Fiction and the depiction of humans in space this is valid from World War I to Vietnam, to the Sputnik-shock to 9/11. Science Fiction has always reflected – and continues to reflect – political, social and technical developments, and last but not least mankind's actual space programmes. Especially World War I and the aerotechnical developments, for instance the establishment of something like an *air-force*, in the years before the actual conflict shaped the form and topics of the phantastic film at the time. While the Brothers Lumière established cinema in 1895 as a form of documentary-based entertainment/edutainment to which we still commit, their counterpart George Méliès, a former illusionist, had completely different plans. The Lumière-clip L'ARRIVE D'UN TRAIN À LA CIOTAT (1895) depicted the arrival of a train in a French station, giving the audience the shock of both modernity and *steampunk*. Two pivotal themes converge in this legendary film: First, we find the idea of *journeying*, which is closely linked to both the dynamics of movement and the frightening origins of cinematography itself. Second – and in no way less terrifying – the clip already embodies the idea of the *fantastic*. In his film VOYAGE DANS LA LUNE (1902) Méliès took this even further: from the platform of the train station to the launch pad of the first cinematic trip to the Moon. In doing so, he not only expanded the narrative limits of the medium film itself, but laid the foundations for the filmic subgenre of Science Fiction – and the narrative use of SFX. As he wrote in 1907: "The substitution trick, called the stop-motion trick, had been discovered and, two days later, I produced the first metamorphoses of men into women and the first sudden disappearances which at first had such great success. [. . .] One trick led to another. Even before this new type was successful, I used my ingenuity to find new techniques and I conceived in turn of the fade obtained by a special device in the photographic camera), appearances, disappearances, metamorphoses obtained by superimpositions on black grounds or on sections set aside in the sets; then came superimpositions on white grounds that had already been exposed (something everyone declared to be impossible before they saw it) realized with the help of a strategem I cannot discuss because imitators have not yet entirely discovered its secret. Then came the trick with cut-off heads, the doubling of characters, of scenes performed by a single character, who through doubling ends by portraying all by himself up to ten similar characters performing a comedy with each other. Finally, using my special knowledge of illusions acquired through twenty-five years of practice at the Théâtre Robert-Houdin, I introduced mechanical, optical, and prestidigitation tricks, etc., to the cinematograph. With all these methods combined and competently used, I do not hesitate to say that in cinematography it is today possible to realize the most impossible and improbable things."[30]

Méliès utilizes the typology of modes of the fantastic described by him above, a mixture of elements from the mostly non-narrative cinema of attractions and the theatre-like parts of story-telling of his times to put the relation between technical development – like space travel – and film as a reflective medium on parallel tracks. The parallel lives of more than 100 years of film and 50 years of space travel led to a strange bond of alternating influence between the two fields. For this paper a selection of 90 film examples, all dealing with human space travel, have been chosen to illustrate the continuities and changes within the history of the genre as well as the history of the before-mentioned relation. Prior to actual space travel, film not only depicted the future, but also reflected the different positions on aviation and coined further developments in this area. As illustrated in the figures of this paper (Figures 7 and 8), films in pre-space travel times mainly depicted mostly destinations within our solar system or at most destinations within the reach of astronomy. Not until the advent of real space travel and the difficult political and cultural climate of the 1950s and 1960s did these subjects almost vanish, returning a decade later with the depiction of invented or unreachable destinations. With the later renewing of manned space travel, planets and stars closer to Earth once again became of interest. This "return" to our solar system also meant, as will be discussed later, the comeback of the (troubled) individual in Science Fiction.

After Méliès it was mainly German utopian films, for instance METROPOLIS (1925/1927) or FRAU IM MOND (1928/1929), and some Russian films, like AELITA (1924), which set new standards for the genre – and for depiction of space travel in general. Beginning with the British production THINGS TO COME (1936) the rethinking of the struggle between ideals of humanity and concepts of

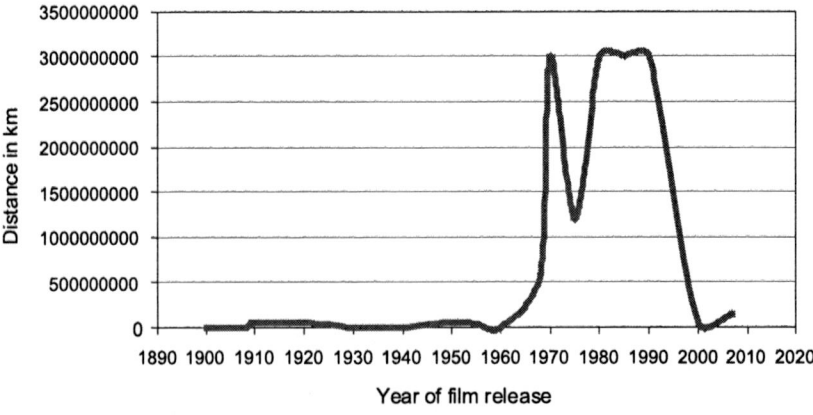

Fig. 7. *Average distance of filmic destinations during a decade (source: Thomas Ballhausen).*

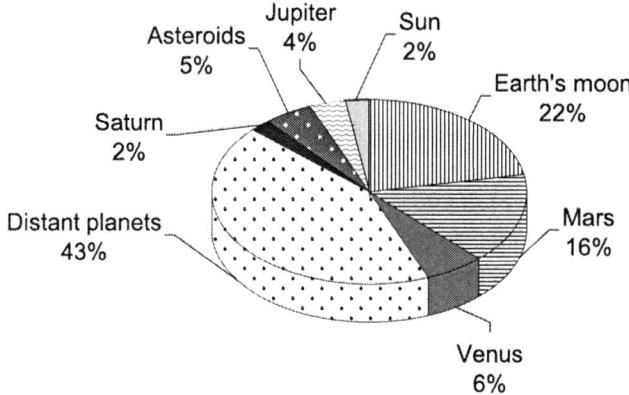

Fig. 8. *Filmic destinations 1902–2007 (source: Thomas Ballhausen).*

technology gained importance again. The essential theme of this British example, namely the subject of hybris, can also be found in the U.S.-American Science Fiction film. Apart from religious aspects, which clearly shaped the early examples, free will and the power to use the given technology were the main issues here until the end of World War II. Especially the so-called *serials* like FLASH GORDON (1936ff.) and BUCK ROGERS (1939ff.) focused on the type of the almost manic go-getter, who rebuilds the universe according to U.S.-American ideals. These aspects can also be found in the space operas of the 1950s, which not only reflect the horrors of war in general, but also particularly those of the Cold War. The real race for space started 1957 with the so-called *Sputnik shock*, and was carried out simultaneously on cinema screens, where the quality of the movies shown reached new heights: "The 1950s marks a turning point in the history of the science fiction film genre. This is a period that is commonly referred to as the 'golden age' of science fiction film, partly due to the unprecedented number of feature films produced and partly due to a group of highly influential, American-made 'classics' released over the course of the decade. [...]".[31] These movies often picked postapocalyptic scenarios as their central theme, depicted the political enemy as the invading alien and reflected (intentionally or not) the fear of a possible nuclear war. Even under the before-mentioned circumstances, the genre of science fiction film still managed to flourish – as did the horror film genre. Both genres dealt with similar issues, but mostly neglecting their full potential, they often focused on matters of the protagonist's moral superiority or the unquestioned use of military power. This tendency – non-reflective fantastic film and the focus on a narrow range of topics – diminished with the beginning of the New Hollywood era in the early 1970s. Mankind then returned to the stars.

1.5.3. Recent examples

Méliès' astronauts in VOYAGE DANS LA LUNE fall asleep shortly after they arrive on the moon. This makes us think about where and especially when we fall asleep, and in which future we might awake. If for example, one wakes up in 1979, humans already travel through cinema-space like truckers. Ridley Scott's ALIEN (1979) not only liberated the depiction of space travel from the camp-aesthetics of the 1950s, but also set it free from the political background of the Cold War era and its narrative imperatives. When the Captain of starship NOSTROMO – nomen est omen – asks the computer in the beginning of the film: "What's the story, mother?", we may answer: "The story in SF, as mentioned before, is everything that's the case." Moreover, ALIEN, though being a more or less classical and non-science-oriented SF-movie, boosted the potential to rethink the value of its genre. The cinematic depiction of space travel became more and more realistic and intertwined with the development of (human) space travel and the destinations within its reach. This trend persists, as can be seen distinctly in contemporary examples, and can only in part be ascribed to SFX and VFX. While 2010: SPACE ODYSSEY (1983), the sequel to Kubrick's classic 2001: A SPACE ODYSSEY (1968), tells of a journey to Io, one of Jupiter's moons, in attempt to find proof for extraterrestial intelligence, the film RED PLANET (2000) heads into another direction. Located in 2050 on an overpopulated Earth, a mission to Mars sets out to save a terraforming-programme previously launched there. The opening sequence of the movie presents an orbital view on our blue planet and is combined with a monologue, which not only sets the mood for the film, but also clearly reflects the positive mainstream arguments and utopian visions for manned space travel to Mars: "By the year 2000, we'd overpopulated, polluted and poisoned our planet faster than we could clean it up. We ignored the problem for as long as we could but were kidding ourselves. By 2025, we knew we were in trouble and began to desperately search for a new home: Mars. For the last 20 years, we've sent unmanned probes with algae bioengineered to grow there and produce oxygen. We'll build ourselves an atmosphere we can breathe. And for 20 years it seemed to work. It looked like we'd pulled it off. Then all of a sudden, oxygen levels started to drop. We don't know why. The international community has thrown all its resources behind us. It's the greatest undertaking man has ever attempted. Our ship, Mars-1, is too massive to launch from the surface of the planet. We've shuttled to the high-orbit space station for a low-gravity launch to begin our six month trip: The first manned mission to Mars. The hope and survival of mankind rests on us. [. . .] We're the first travellers to another planet. It's another giant leap for mankind and if we don't find out what's wrong on Mars, it could be our last."[32]

Another crew, this one consisting of the protagonists of SUNSHINE (2007), and their spaceship Icarus II, are backed up by all nations of Earth. The ship is carrying an enormous payload: a bomb the size of Manhatten, with which they intend to reanimate the dying sun. Again, a manned mission seems to be the only way to save mankind. However, this movie avoids all stereotypes of the genre: "In Danny Boyle's *Sunshine* we do not see people gathering in bars to watch newcasts from Paris or London and cheering on the heroic astronauts of Icarus II on their mission to revive the sun. When at last we get a glimpse of Earth, stuck in near-permanent winter, it's a rather unapocalyptic Christmas-card landscape with kids making snowmen. No one whoops it up at mission control, and for all we know the only person who realises the mission has succeeded in 'saving the mankind and so on' is the sister of the ship's chief physicist Capa, who receives a video message from him. By eliminating the disaster-movie convention of giving doomsday an on-screen audience, as in the generically similar *Armageddon* (1998) and *The Day after Tomorrow* (2004), Sunshine removes the sickly screen-traversing sense of camarderie in the face of catastrophe that – along with the spectacle of cities being destroyed – forms a part of those films' appeal."[33] Besides looking at dimensions of size, for example the spatial relation between the ship and the solar system or the crew's quarters and the size of the bomb, etc., SUNSHINE is far more realistic in depicting the dangers and unavoidable sacrifices of human space travel. With films like SUNSHINE or THE FOUNTAIN (2006) the journey to the stars, to destinations relatively close to Earth or distant planets likewise, has been turning more and more into an expedition into the depths of the travelling individuals, a movement towards the "I". To rethink and reformulate Darko Suvin's definition of Science Fiction literature: these new Science Fiction films about cinematic space travel whose destinations are, in fact, of interest for actual space travel, are part of a genre of cognitive entrangement, which forces the audience to look at a partly defamiliarised reality, thus encouraging it to contemplate upon the known world from a more distanced perspective.[34]

So, what are we going to find in outer space? According to recent examples we will, presumably find ourselves, besides all wonders. Throughout film history the subject of human space travel has always been an integral part of science fiction's narrative conventions. Albeit being highly influenced by concomitant political, cultural and technological circumstances, the genre has also invariably been an expression of these conditions and will continue to be, even in the case of a majority of fantastic films temporarily dealing with subjects other than space travel itself, in response to actual political situations.

1.5.4. Appendix: list of evaluated film examples

1. LE VOYAGE DANS LA LUNE (FR 1902), Dir.: Georges Méliès
2. LE VOYAGE SUR JUPITER – UN EXCURSION CHEZ JUPITER (FR/ES 1909), Dir.: Segunde de Chomón

3· HIMMELSKIBET – THE SHIPS OF HEAVEN (DK 1918), Dir.: Holger Madsen
4· HELLO MARS (USA 1923), Dir.: Alfred J. Goulding
5· AELITA (UDSSR 1924), Dir.: Iakov Protasanoff
6· FRAU IM MOND (DE 1928/1929), Dir.: Fritz Lang
7· DESTINATION MOON (USA 1950), Dir.: Irving Pichel
8· FLIGHT TO MARS (USA 1951), Dir.: Leslie Selander
9· WHEN WORLDS COLLIDE (USA 1951), Dir.: Rudolph Matés
10· RED PLANET MARS (USA 1952), Dir.: Harry Horner
11· ABBOTT UND COSTELLO GO TO MARS (USA 1953), Dir.: Charles Lamont
12· SPACEWAYS (GB 1953), Dir.: Terence Fisher
13· RIDERS TO THE STARS (USA 1954), Dir.: Richard Carlson
14· CONQUEST OF SPACE (USA 1955), Dir.: Byron Haskin
15· FORBIDDEN PLANET (USA 1955), Dir.: Fred McLeod Wilcox
16· MESSLE TO THE MOON (USA 1958), Dir.: Richard E. Cunham
17· NEBO ZOVYOT (UDSSR 1959), Dir.: Alexander Kosyr, Michail Karjukow
18· DER SCHWEIGENDE STERN (PL/DDR 1959/1960), Dir.: Kurt Maetzig
19· THE ANGRY RED PLANET (USA 1959/1960), Dir.: Ib Melchior
20· 12 TO THE MOON (USA 1960), Dir.: David Bradley
21· UCHU DAI SENSO – BATTLE IN OUTER SPACE (JP 1960), Dir.: Inoshirô Honda
22· THE PHANTOM PLANET (USA 1961), Dir.: William Marshall
23· PLANETA BUR (UDSSR 1961), Dir.: Pawel Kluschanzew
24· YOSEI GORATH (JP 1962), Dir.: Inoshirô Honda
25· IKIARIE XB1 – IKARIE XB1 (CZ 1963), Dir.: Jindrich Polák
26· ROBINSON CRUSOE ON MARS (USA 1964), Dir.: Byron Haskin
27· FIRST MEN IN THE MOON (GB 1963/64), Dir.: Nathan H. Juran
28· 2071 – MUTAN – BESTIEN GEGEN ROBOTER – THE TIME TRAVELLERS (USA 1964), Dir.: Ib Melchior
29· IL PIANETA ERRANTE (IT 1966), Dir.: Antonio Margheriti
30· TERRORE NELLO SPAZIO (IT/ES 1965), Dir.: Mario Bava
31· SS-X-7 (USA 1965), Dir.: Hugo Grimaldi
32· COUNTDOWN: START ZUM MOND (USA 1966), Dir.: Robert Altman
33· WAY...WAY OUT (USA 1966), Dir.: Gordon Douglas
34· PERRY RHODAN (IT/BRD/ES 1966/1967), Dir.: Primo Zeglio
35· ROCKET TO THE MOON (GB 1967), Dir.: Don Sharp
36· TUM ANOST ANDROMEDY (UDSSR 1967), Dir.: Yevgeni Sherstobitov
37· BARBARELLA (FR/IT 1968), Dir.: Roger Vadim
38· VOYAGE TO THE PLANET OF PREHISTORIC WOMEN (USA 1968), Dir.: Peter Bogdanovich
39· 2001 – A SPACE ODYSSEY (GB/USA 1968), Dir.: Stanley Kubrick
40· MAROONED (USA 1969), Dir.: John Sturges
41· TO THE FAR SIDE OF THE SUN (GB 1969), Dir.: Robert Parrish
42· SIGNALE – EIN WELTRAUMABENTEUER (SIGNALY-MMLX) (DDR/PL 1969/197), Dir.: Gottfried Kolditz
43· HORROR OF THE BLOOD MONSTERS (USA 1970), Dir.: Al Adamson
44· SILENT RUNNING (USA 1971), Dir.: Douglas Trumbull
45· ELOMEA (DDR 1971/1972), Dir.: Herrmann Zschoche
46· SOLYARIA (UDSSR 1972), Dir.: Andrei Tarkovsky
47· UKROSHENIYE OGNYA (UDSSR 1972), Dir.: Danil Kurabrovitsky
48· MOSKWA-KASSIOPEIJA (UDSSR 1973), Dir.: Richard Wiktorow
49· DARK STAR (USA 1974), Dir.: John Carpenter
50· THE DOOMSDAY MACHINE (USA 1974), Dir.: Lee Sholem
51· STOWAWAY TO THE MOON (USA 1974), Dir.: Andrew V. McLagen

52· DIE PHANTASTISCHE WELT DES MATHEW MADSEN (BRD 1974), Dir.: Helmut Herbst
53· BLACK SUN (GB/IT 1976), Dir:: Lee H. Katzin, Ray Austin
54· OPERATION GANYMED (BRD 1977), Dir.: Rainer Erler
55· TEST PILOTA PIRXA (PL/UDSSR 1978), Dir.: Marek Piestrak
56· ALIEN (GB/USA 1979), Dir.: Ridley Scott
57· THE MARTIAN CHRONICLES (GB/USA/BRD 1980), Dir.: Michael Anderson
58· PETLJA ORION-PHAETON AN ERDE (UDSSR 1980), Dir.: Wassili Lewin
59· GALAXY OF TERROR (USA 1980), Dir.: Bruce Clark
60· TAYNA TRETEY PLANETY (UDSSR 1981/1982), Dir.: R. Kacioanov
61· FORBIDDEN WORLD (USA 1982); Dir.: Allan Holzman
62· THE RIGHT STUFF (USA 1983), Dir.: Philip Kaufman
63· SPACEHUNTER: ADVENTURES IN THE FORBIDDEN ZONE (USA 1983), Dir.: Lamont Johnson
64· EXPLORERS (USA 1984), Dir.: Joe Dante
65· SAYÔNARA SIYÛPETÂ (JP 1984), Dir.: Koji Hashimoto, Sakyo Komatsu
66· WOSWRASCHTSCHENIJE S ORBITY (UDSSR 1984), Dir.: Alexander Surin
67· ALIENS (USA 1985), Dir.: James Cameron
68· CREATURES (USA 1985), Dir.: William Malone
69· GA, GA – CHAWALA BOHATEROM (PL 1986), Dir.: Piotr Szulkin
70· ÔRITSU UCHÛGUN ONEMAISO NO TSUBASA – THE WINGS OF HONNEAMISE (JP 1987), Dir.: Hiroyki Yamaga
71· EARTH STAR VOYAGER (USA 1988), Dir.: James Goldstone
72· SOLAR CRISIS (JP/USA 1990), Dir.: Richard C. Sarafian
73· STAR VOYAGER (USA 1994), Dir.: Rick Jacobson
74· APOLLO13 (USA 1995), Dir.: Ron Howard
75· THE COLD EQUATIONS – EMERGENCY IN SPACE (CA 1997), Dir.: Peter Geiger
76· FALLING FIRE (CA 1997), Dir.: Daniel D'Or
77· ROCKETMAN-SPACEMAN (USA 1997), Dir.: Stuart Gillard
78· STARSHIP TROOPERS (USA 1997), Dir.: Paul Verhoeven
79· CONTACT (USA 1997), Dir.: Robert Zemeckis
80· ARMAGEDDON (USA 1998), Dir.: Michael Bay
81· LOST IN SPACE (USA 1998), Dir.: Stephen Hopkins
82· GALAXY QUEST (USA 1999), Dir.: Dean Parisot
83· PITCH BLACK (AUS/USA 1999), Dir.: David N. Twohy
84· MISSION TO MARS (USA 2000), Dir.: Brian DePalma
85· RED PLANET (USA/AUS 2000), Dir.: Anthony Hoffman
86· SPACE COWBOYS (USA 2000), Dir.: Clint Eastwood
87· SOLARIS (USA 2002), Dir.: Steven Soderbergh
88· STRANDED: NÓ UFRAGOS (ES 2002), Dir.: Mária Lidón
89· PER ANHALTER DURCH DIE GALAXIS (USA/GB 2005), Dir.: Garth Jennings
90· SUNSHINE (GB 2007), Dir.: Danny Boyle

---

[30] Georges Méliès: Trick Effects. In: The Science Fiction Film Reader. Edited by Gregg Rockman. New York: Limelight Editions, 2004. pp. 2–4, p. 2ff.

[31] Christine Cornea: Science Fiction Cinema. Between Fantasy and Reality. New Brunswick, NJ: Rutgers University Press, 2007. p. 30.

[32] RED PLANET (2000), DVD-edition Warner Bros. 2001, Timecode 00:00:30-00:03:11.

[33] Henry K. Miller: Sunshine. In: Sight & Sound Iss. 4 (Vol. 17) 2007. pp. 81–82.

[34] Cf. Christine Cornea: Science Fiction Cinema. Between Fantasy and Reality. New Brunswick, NJ: Rutgers University Press, 2007. pp. 2–11.

1.6 Aiming ahead: next generation visions for the next 50 years in space

Agnieszka Lukaszczyk

1.6.1. The past and the future

The year 2007 has been a very important one for space; it marked the 50th anniversary of the Space Age. Such a moment calls for a reflection on the past and a review of accomplishments to date, but moreover, it calls for a planning of the future, which may take humanity to places that past generations could only dream about. Throughout the last half of the century, humanity has achieved great advances in many different areas. Technological endeavour has been accompanied by a tremendous social, cultural and economic impact. Space activities have entered an era where a mixture of various disciplines from sciences to humanities will be necessary for future space exploration enterprises. This raises the question of what will happen next? Or rather, what should happen next?

The Space Generation Advisory Council (SGAC) in partnership with the Frederick S. Pardee Center for the Study of the Longer-Range Future, the Boston University Center for Space Physics, The Planetary Society and the Secure World Foundation, has conducted an extensive project to determine youth visions for the next 50 years of space exploration. The first stage of the project combined 900 youth visions supplied in an online survey, which were summarised by a team of youth volunteers and presented at the symposium "The future of space exploration: Solutions to earthly problems" in Boston, U.S. that took place from 12 to 14 April 2007 featuring prominent individuals such as Freeman Dyson and Dr. Harrison Schmitt. Questions that have emerged in the survey included but were not limited to:

- What will space look like in the next 50 years?
- How can we learn from our history on earth as we move into this next frontier?
- How can we create opportunities for the sustainable, beneficial and effective use of space?
- Who will decide these questions?
- How will conflict of interests be settled?
- What new systems, structures and paradigms do we need as we begin this new adventure?[35]

1.6.2. Aiming ahead

It has been fascinating to study the desires, dreams, hopes, and plans of today's youth for the next 50 years. The next generation has been quite aware of the possible opportunity to explore new places and discover new worlds. *Those adventures will be driven by the human quest for knowledge and human curiosity. They will provide the main opportunity for equitable international cooperation. Humans divided on Earth will hopefully unite in space as citizens of one planet.*[36] Moreover, an interplanetary society may emerge at the moment a first child is born on another planet.

Many of the ideas that resulted from the survey have been more the product of dreamers than of realists. However, one should remember that similar dreams have lead to incredible discoveries. One hundred years ago, no one would have believed that man could walk on the moon. Thus, although the results of the survey may raise a brow or two, they should not be taken lightly, as we have to remember that the young people of today are the leaders of tomorrow.

Around 700 youths from the ages of 18–29 were surveyed worldwide. The questions asked were open-ended, as the goal was to allow the imagination to flourish without restrictions and without leading the respondents to certain answers.

1.6.3. Ensuring the survival interests of humanity

1.6.3.1. The sustained exploration of the Moon

There was an overwhelming response in favour of exploring the Moon, using it as a test bed for the exploration of other solar system bodies, for the development of a

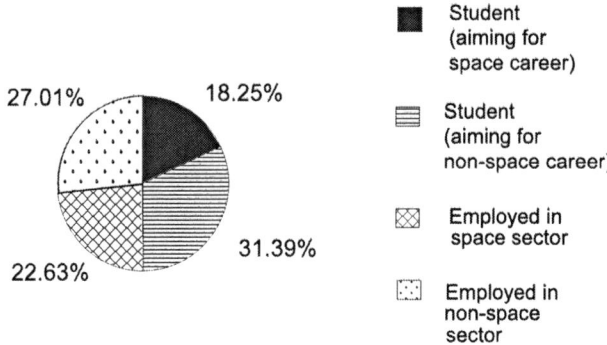

Fig. 9. *Division of the survey participants (source: results of the survey).*

45

permanent moon base or bases, and the development of a cis-lunar economy based on the extraction of different resources in order to make exploration cost-effective. A large majority of the responses noted that the first priority of resource use on the moon should remain with providing the means to sustain life supplies as well as the well-being of the lunar explorers. Respondents felt strongly that before humans explore further planets, they should be able to sustain themselves on the moon for significant periods of time. Recognizing the need for a self-sustaining lunar settlement, research and the development of efficient sustainable food, water, and air production/recycling systems should be made a priority.

1.6.3.2. Habitat and life sciences

Respondents recognized the unique benefits afforded by making exploration and long-term human presence in outer space major global objectives. They supported the pursuit of technologies that would enable long-duration habitats on the moon and Mars. They recommended the creation of international standards for space habitat construction, including but not limited to interfaces between modules of different national origin; occupational health and safety; and a module of structural design. They proposed fixing a universal maximum radiation exposure limit to protect space inhabitants as well as to support international collaboration for space habitat development. In addition, respondents supported habitat design by diverse technical teams, which included engineering and social sciences, as well as other relevant disciplines. They also encourage regulations that protect the general public, while not imposing undue hardship on entrepreneurial space efforts and which facilitate an active dialogue for creating provisions for real property rights in outer space. In addition, respondents recognized the risks inherent to space flight. They expressed that in order to sustain a human presence outside Earth, a better understanding of human physiology in low-gravity for extended periods of time is needed. In light of the American Vision for Space Exploration, they urged an intensification of human physiology research on ISS; suggested the additional development of biosatellites similar to the Mars Gravity Biosatellite architecture pioneered by the teams at MIT to study physiology on short-term gestation mammals in orbit, simulating lunar and Martian gravity, and returning them safely to Earth.

1.6.3.3. Furthering exploration of the solar system

Sustained exploration of the moon as well as continual human presence in near-Earth orbit should be used to yield new and improved technologies that can help foster the further exploration of the planetary bodies. When asked which technologies should be given the highest priorities, the majority (49%) felt that life support systems that work longer and more efficiently would be beneficial for Mars exploration. There is also a strong mandate for exploring Mars in support of extending science research to allow us to better understand our existence in a three-body system.

1.6.3.4. Maintaining human presence in the near-Earth space

There is strong support for completion of the International Space Station prior to a reassessment in order to maximize its use. However, it seems that a large percentage of the respondents were not convinced that this would happen, as 54% of them felt that there is a need for a new station to be built cooperatively by governments and private industry. This human outpost can be used as a docking station for further missions (38%), to maximize science benefits and attain cost-effectiveness to carry out exploration (34%) and to pay more attention to safety and reliability.

1.6.3.5. Space governance

Space efforts need to be led by all nations collectively including space-faring countries under the umbrella of international organizations such as the United Nations. Apart from some very important revisions to the Outer Space Treaty (OST), there is a need to include the registration of defence-related missions in order to reduce the secrecy and security threats that still surround businesses within the space environment. It was also noted that important revisions to the OST are needed, in particular, revisions pertaining to the destinations of near-term exploration such as cis-lunar, moon surface and Mars under this new governance. There was strong support for internationally cooperative efforts for space exploration and to help humanity benefit from the coordination of research efforts from all countries.

1.6.3.6. Peaceful uses of outer space

It was very apparent in the survey that it was of a crucial importance to the respondents to make outer space secure and used only for peaceful purposes. They recommended the space agencies and governments to continue their dialogue on the future military uses of space in order to establish rules, which would benefit society in

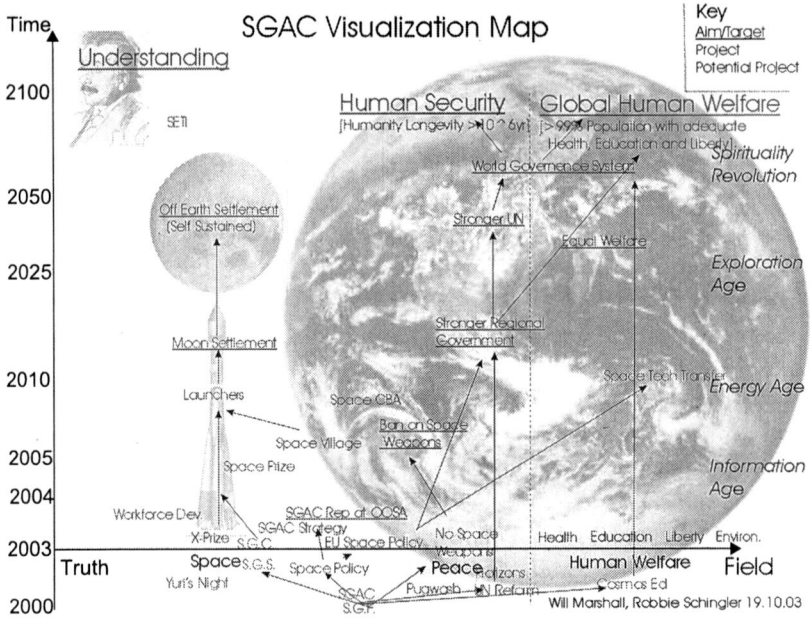

Fig. 10. *SGAC Visualization Map (source: Will Marshal. Robbie Schingler, Space Generations Advisory Council).*

general. When thinking about outer space, young people no longer see each other as citizens of individual countries, but as citizens of the world. Rather than competition among nations, they favour cooperation at the international level with a deep understanding of the necessity for global collaboration in outer space. In addition, the respondents felt strongly about the Disaster Monitoring Constellation in the sense that it is a globally coordinated, cooperative effort to protect Earth. They would like to see an international effort to create a 'global network of observation' to encourage the free flow of information to subvert the potential threat of global disasters and redouble efforts to observe all large near-Earth objects (NEOs) by 2010. We encouraged the increase in public awareness regarding the potential for current and developing space technologies created for the purpose of averting potential disasters originating from Earth or from the threat of near-Earth objects.

1.6.3.7. Private use of space

Many respondents have been inspired by the recent developments in the private space sector, including the space flights with private space tourists aboard. They supported the progress of private space industry, because the shared belief was that outer space should be more inclusive and thus accessible to ordinary people. Many ordinary people consider use of space as being reserved only for the space-faring nations. When surveyed, 81% of respondents in Europe (outside of the space sector) did not even know that there is a European Space Agency in Europe. This reveals that the awareness of space matters is very low. Although people use space applications on daily basis (telecommunications, weather forecasts, navigation, and more) they do not connect the two. When asked about outer space, most feel uncomfortable, as the topic is very foreign to them. Hence, respondents believe that by making the use of space more accessible, average citizens would have a better chance to become acquainted with the wonders of the skies above.

1.6.3.8. The role of developing countries in space exploration

As mentioned before, outer space may serve as a unique platform for the citizens of the world to collaborate in various endeavours. Hence, the benefits of space technologies should be accessible to all. Today, we are witnessing a resurgence of interest in lunar exploration. Worldwide, the motivation of space agencies is increasing to spend more time and money on lunar research activities. Improving the practicality and reliability of advanced technologies in space transportation is crucial for the enduring exploration of space. It is important to recognize the economic opportunity that has inspired previous exploration ventures, and remember this for future exploration strategies. Reports and studies such as the 2007 International Lunar Decade by the Planetary Society promote international collaboration to contribute to technological progress. This cooperation can be expanded to encourage developing countries to become involved in space programmes. Respondents encouraged fostering capacity-building between countries, intergovernmental organizations and/or NGOs. They proposed achieving this by advancing space capabilities within developing countries. They encouraged developing nations to pool resources and get involved in space-related operations by promoting education and research, and creating an infrastructure to permit the development of a potential space programme. Awareness and information on space exploration must be encouraged in the developing nations.

1.6.3.9. Space technology development

Respondents recognized the importance of public support as a major driver for the success of space programmes, particularly in the fields of medicine, agriculture,

public safety, and disaster management, in addition to recognizing the limited
informational outreach of space technologies by federal agencies and commercial
space industry. Furthermore, they recognized the ability of humankind to become a
space-faring civilization, and bearing in mind the importance of promoting space
among developing countries and the need to involve them in the space sector
workforce, respondents believed there should be additional funding for the research
and investigation of advanced and breakthrough space transportation technologies.
There should be a revival of aeronautical research in both private and public sectors.
Hypersonic research programs are positive initiatives in aeronautical technology
development. This should continue because such technology would benefit applications including space access, commercial airline transportation, and future
supersonic transports. There should also be continued research and application
of interplanetary superhighway trajectories for in-space transportation. They
recommended the creation of information flows to developing countries focused
on space science and space technologies to raise the technological base of those
countries, and to further the development of a space industry. The importance of
elaborating faster and more efficient export control procedures in order to increase
the competitiveness of all markets related to the space sector has been emphasized.
Taking into account the low failure rate of human space transportation systems, the
restrictive costs of payload to Low Earth Orbit, and recognizing the alarming
decline in investment in research of technologies, they recommended that research
and development be conducted in the areas of control, thermal management and
propulsion technologies, particularly on pulse detonation engines, hypersonic

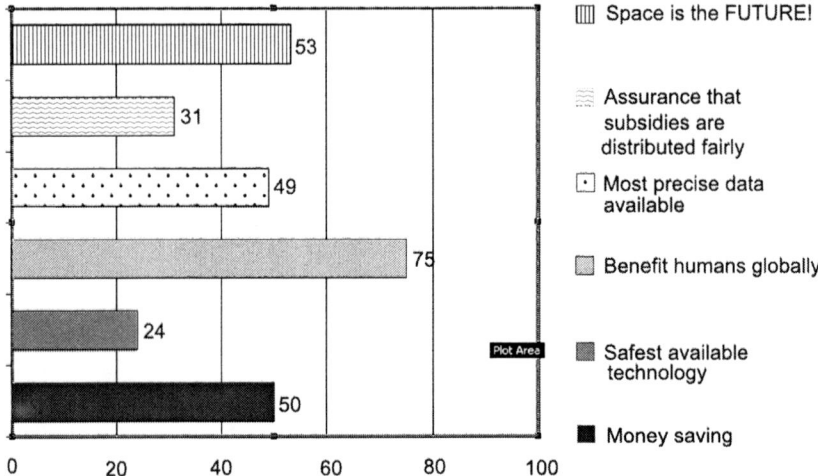

Fig. 11. *Political incentives for developing space applications (source: Agnieszka Lukaszczyk).*

engines, and electric propulsion. In addition, respondents supported investments into the research and development of space-related technologies, that could eventually lead to more efficient operations, with the objective of decreasing the cost of travel to Low Earth Orbit by one order of magnitude.[37]

1.6.3.10. Space education and outreach

There is a lack of space education in schools, especially in developing countries. Students are not aware of the opportunities that exist in fields such as remote sensing applications, satellite communication, or future space programmes in their country. Through a more comprehensive space education, they will be able to apply this knowledge to solve specific problems within their community, thus contributing to the economic development of their nation. Creation of a Global Space Education Curriculum – expanding space education in schools on an international level by convincing governments and schools to include space curriculum in classrooms has been recommended. These programs will raise space awareness as well as stimulate student interest in studying science and engineering. Organizations such as UNESCO, among others, should play an active role in encouraging educational programmes in space research. Space agencies should regularly inform the public, especially people from developing countries, of the benefits of space technologies by setting up specific workshops and educational events. Young people pointed out that it appears as if many agencies, organizations and governments are aware of the poor information distribution when it comes to space topics, however, very little progress has been achieved in this area.

1.6.4. Conclusions

This survey may be viewed as a testimony of young people on space issues. Often, those in power do not take the opinions of youth into consideration due to the assumption that today young people have nothing relevant to say about subjects of high political importance. Fortunately, studies such as this one prove quite the contrary. Young people are not only engaged in issues such as space matters but also want to be heard. Their enthusiasm and sometimes naïve yet pure desire for a better tomorrow for all humankind should serve as an example for decision makers. No one expects his or her recommendations to be put into practice immediately. However, a careful evaluation of this global voice would be advisable.

After all, young people represent the future of the space industry. Their recommendations combine the idealism and vision of youth with the realism gained from the first steps within the space sector during their studies and employment. As a link to the next generation, they provide proposals aimed at helping to check the falling number of students of science and engineering in tertiary-level education. Moreover, they provided a framework of recommendations that they believe can enable the best use of space to the benefit humankind.[38]

One must remember that without dreams and ideals progress is limited. It sometimes helps to "step outside of the box" and listen to the whispers. Such whispers, when combined, make a powerful voice – a voice that should not be neglected. Most assert that the outreach and education regarding space matters has not been adequate. The reassuring thing is that there are young minds out there willing to work very hard to promote space and make the best use of it. Such human resources should not be disregarded. More than that, such human resources should be taken advantage of before they change their minds and go into other and more rewarding fields.

"Shoot for the moon, even if you miss it you will land among the stars."

Unknown

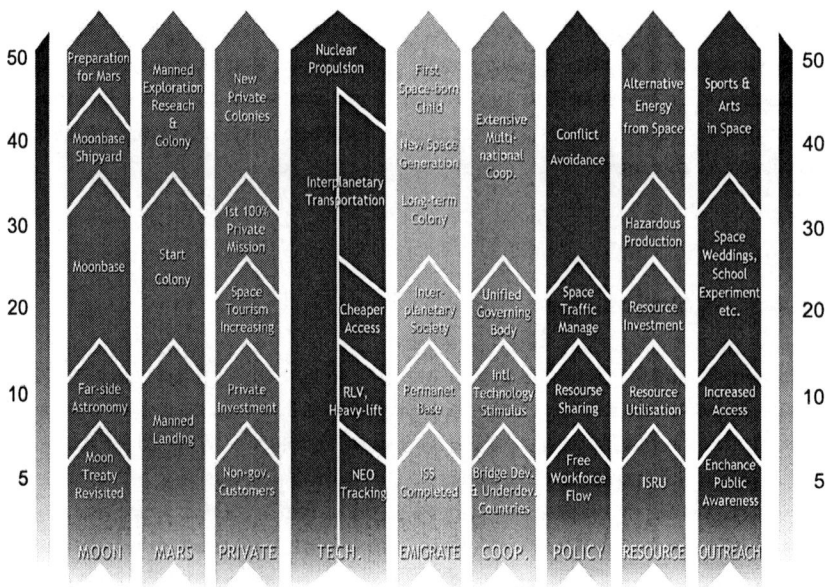

Fig. 12. *SGAC 50-Year vision roadmap (source: Space Generations Advisory Council).*

[35] "50 Year Visions for Space Exploration. SGAC Research Team". SGAC. http://www.spacegeneration.org/node/254.

[36] See. "The Vienna Vision". pp 229–233. In this volume.

[37] Space Generation Congress. Youth Declaration. Valencia, Spain, 2006.

[38] Thakore Bee, ed., Boshuizen Chris, Firestone Tiffany, Marshall William, Schingler Robbie, Shi H (2007) Vision for the Next 50 Years of Space Exploration on the occasion of the 50th Anniversary of Space Exploration Recommendations from Students and Young Space Professionals, Report to UN COPUOS, Space Generation.

CHAPTER 2

CAN WE COMPARE?

2.1 Summary

Monique van Donzel

Prior to embarking on the odysseys which developed into the Vienna Vision on Humans in Outer Space, a series of papers address human exploration against a historical and anthropological background of exploration as an inherent human activity. Can we compare, and can analogies be used? The contributions to this section set the scene for the three odysseys in the following sections. The contributions each focus in their specific way on the role and situation of humans around the Earth, their place in exploration, and the search for life in the universe. This section addresses the questions set out in the general introduction to this volume: Should humans explore space? Do the (cultural and economic) drivers for exploration require human participation? What are the human abilities and reasons to adapt to such extreme conditions as presented by the space environment beyond Earth? Should man be prepared for – ethical and societal – consequences of an encounter with Extraterrestrial life?

The contribution of James Muldoon focuses on the question of how historical situations can provide us with analogies that would help us understand the problems associated with space travel and the regulation of space. He presents a useful analogy from the 16th century with the discovery of the New World and the subsequent debate about access to and regulation of the Ocean Sea and the development of a legal regime for the seas. This is analogous to the current age of space exploration and regulation. Sixteenth century explorers, standing on the edge of the Ocean Sea, faced a similar situation as a modern space traveler on the launch pad at Cape Canaveral. Modern governments will have to develop rules of access to and activities in space, in order to seek the elimination of potential sources of conflict in space. In doing this, legislators, judges and lawyers will be retracing some of the steps that their predecessors trod in the 16th and 17th century.

Gísli Pálsson addresses the issue of exploration from the human and within the human. Now that most of the habitat of the globe has been charted, documented and conquered, humans are increasingly turning their attention to the 'remote corners' of both outer space and the living organism, in particular the human genome. Focusing on languages of voyaging, spatial representation, procedures of mapping, and the development of property regimes, his paper discusses the similarities and differences of the exploration of outer space on the one hand,

57

and research on what molecular biologists sometimes refer to as the 'universe within' – the human genome – on the other. These issues are central to the third odyssey.

The debate about humans as opposed to robots in space is reflected clearly in the kind of tasks entrusted to humans and robotic devices, respectively, in the two leading space powers of the early space age, the U.S. and Russia. This is the focus of the contribution by Sven Grahn. Analogies with for instance aeronautics often drove the debate on how to allocate tasks between humans and machines. Grahn's contribution shows that analogies, as useful as they may be, should also be used with caution. An analysis of each of the analogies used in the past to justify human spaceflight, shows that they all have weaknesses. One should be careful in using such analogies when charting the course of humans into space. Space is different. The ideas and suggestions from this contribution are also relevant in the first odyssey.

2.2 *Inter caetera* and outer space: some rules of engagement

James Muldoon

2.2.1. Introduction

On 19 January 2007 the issues with which this conference is concerned suddenly became front page news instead of the stuff of science fiction – when the world became aware that the Chinese government had employed one of its medium-range ballistic missiles to shoot down one of its own outdated weather satellites, presumably to demonstrate that government's interest in the future of space travel. The actions of the Chinese government serve to remind us that while space is vast and there would seem to be room for any state interested in space exploration and exploitation, in fact such is not the case. Although the writers of fiction carry us imaginatively out into distant galaxies, traveling at speeds vastly faster than the speed of light, the realities of space travel keep us closer to earth, at least for the foreseeable future.

In a loose sense, the space above and around us consists of three zones, each with its own characteristics and problems. The closest zone is that associated with airplane travel, a region that a number of international conventions regulate and that are ultimately enforced by the threat of not being granted landing rights if an airline does not adhere to the regulatory regime. Weather, communications, and military satellites that circle the earth in ever increasing numbers along with ever greater amounts of space junk occupy the second zone. For our purposes, let us assume that this layer extends to but not beyond the moon. This region is presently unregulated in any formal way, although self-interest requires that those sending satellites into space ensure that their satellites do not interfere with the orbits of other satellites. The Chinese destruction of one of their own satellites brought renewed calls to demilitarize space, an action that would require some sort of unified effort on the part of the major nations of the earth if it is to be effective. The third zone of space is the universe beyond the moon, a region where there is little actual activity at the moment and there is little evidence that there will be any in the immediate future.

Each of the three zones of space presents two fundamental problems that at some point have to be addressed. The first problem is access. Who has the right to enter each space, any individual, group, corporation, or other private entity, or should access to space be restricted to states? The second issue is the regulation of activity in space. Can there be or should there be regulation and, if so, how will regulations be determined and enforced?

Let me make one more assumption for the thesis of this paper: the cost of entering space will decline to the point that small nations, alone or in concert, and wealthy private entities could afford to enter the space race. If the cost declines sufficiently, there could be something approaching traffic snarls in those orbits that are most useful for communications, weather, and military satellites. Furthermore, as more satellites are sent up to replace obsolete ones, as others come to the end of their effective life, and as older satellites begin to fall apart, there is the problem of space junk cluttering up the heavens and eventually causing collisions. These issues have implications for a range of activities from military planning to insurance against the dangers in space. Attempts to regulate this zone lie in the foreseeable future. At the moment, however, about the only regulation arises from cost/benefit considerations. Is it worthwhile to spend money on further space exploration?

2.2.2. Regulating the Ocean Sea

Given these assumptions, do we have any historical situations that provide us with analogies that would help us understand the problems associated with space travel and regulation or are these problems absolutely unique? As the title of this paper suggests, in fact we do have an analogous situation, the discovery of the New World, the subsequent debate about access to and regulation of the Ocean Sea, that is the Atlantic Ocean, and the development of a legal regime for the seas. In 1493, following Columbus's return from his first voyage, Pope Alexander VI (1492–1503) issued three bulls generally known by the name of the first, *Inter caetera*, that were a significant attempt to impose European Christian principles of legal order on the Ocean Sea and to the lands beyond. Within the obvious limits that any historical analogy present, the debate that emerged about whether the sea was open, *mare liberum*, or closed, *mare clausum*, provides us with some interesting parallels to the current debate about the regulation of space. Furthermore, *Inter caetera* and related papal and canonistic documents made an important contribution to the 16th- and 17th-century efforts to create international law, a process that reached its peak with the publication of Hugo Grotius's *On the Law of War and Peace* in 1625, the culmination of two centuries of legal discussion about access to and jurisdiction over the sea that had begun with *Inter caetera*.

As space presents us with three zones of interest, so too there were three zones of the sea that concerned early modern rulers and their lawyers. The first zone consisted of coastal waters, the second of large areas of the adjoining seas that some states claimed to possess or at least to possess jurisdiction over, and the third was the Ocean Sea. The first led to such conflicts as that between the English and Scots on the one hand and the Dutch on the other about access to the herring fishery in the North Sea. Could the English and Scots legitimately keep the Dutch and other outsiders from these fishing grounds?[39] By the 18th century this kind of conflict led to the creation of the three-mile rule, that is, that a nation's jurisdiction extended to a point three miles off shore but no further. This rule, like the regulation of civil aircraft, was effective and enforceable, because it was in the mutual self-interest of seafaring nations.

The second zone concerned the claims of the Venetians to 'own' the Adriatic and the Genoese to 'own' the Ligurian Sea. These states claimed the right to charge tolls and otherwise exercise jurisdiction over those who sailed in these seas. In effect, these and similar claims by other nations asserted that the sea could be closed, a *mare clausum* as the Portuguese legal theorist Serafim de Freitas argued, in defending the Portuguese claim to possess a monopoly of trade between Asia and Europe.[40]

The third zone was the Ocean Sea, the extent of which was unknown for a long time. The Ocean Sea was a highway to the wealth, real or imagined, of as yet unreached lands of which Europeans had heard and whose products they desired. As Europeans were to learn, it also contained lands the existence of which they were completely unaware. Could any nation or group of nations lay claim to the Ocean Sea and therefore to the routes that provided access to that wealth and to the potentialities of the newly discovered lands?

These were not entirely new questions. Canon lawyers had been interested in such questions since the mid-13th century and there existed a small body of legal theory and a number of papal documents that addressed some of the basic issues that arose after 1492, underlying what Lewis Hanke described as The Spanish Struggle for Justice in the Conquest of America.[41] *Inter caetera* was thus not a radical assertion of some new kind of papal power but rather the application of a set of 2-centuries old legal principles to the circumstances that attended Columbus's first voyage, principles that several popes had already employed during the 15th century to deal with access to the Atlantic islands. The basis for such a papal role was the theory of universal papal jurisdiction that the canonists had developed. Underlying this legal structure were several fundamental assumptions about mankind. The first was that all mankind descended from Adam and Eve and therefore ultimately formed a single people with a common origin and purpose. The second was that mankind was flawed, all men being subject to the consequences of Adam's original sin and therefore the beneficiaries of Christ's redemptive sacrifice. Wrestling with Christian theology and Greek philosophy, the medieval scholastics had developed the theory of the dual nature of mankind, natural and supernatural, each with its own end. Within this framework, humanity had both a supernatural end and also a natural one, each structured in terms of law, eternal law and that part of the eternal law accessible by human reason, that is the natural law. Each of these laws was reflected in the positive laws of human societies, canon law for the universal Church and the various kinds of secular law established for specific societies. In ecclesiastical legal and political thought there was also a tendency to believe that in an ideal world mankind would form a single organized society under a single head. This concept resonated with the notion of a Christian Roman Empire and universal Roman Law that tantalized a number of medieval intellectuals. Some theorists argued for a dual headship of Christian society, pope and emperor, but others denied this argument.[42] While the Christian emperor could be seen as the superior of all other Christian rulers in some hierarchical order, the emperor in turn was subject to the pope in spiritual matters, so the pope, who represented God whose eternal law was the basis of all other law, stood at the apex of the human legal hierarchy.

2.2.3. The universal human community

The canonist Sinibaldo Fieschi, better known as Pope Innocent IV (1243–1254), applied the legal principles that the scholastics had developed to human society in concrete terms. He argued in effect that all mankind is subject to one of three legal orders, canon, Mosaic, or natural, and the pope is the supreme judge of each legal realm. The papal role in relations between Christian nations is obviously the easiest claim to accept.[43] In a variety of ways popes had been attempting with mixed success to regulate international affairs within Latin Christendom for several centuries. This claim was an important element of the larger Church-State issue that characterized medieval society. The claim that the pope was authorized to intervene in Jewish communities within Christian Europe when they were not adhering to the Mosaic Law is more difficult for us to accept. From the papal perspective, however, it was a part of the papal responsibility for the protection of the Jewish people because of their preparatory role in God's revelation of Himself to mankind. Errors had crept into the Jewish tradition by way of the Talmud, and it was the responsibility of the pope to insure the doctrinal purity of the Mosaic tradition.[44]

The final stage of the pope's jurisdiction, as ultimate judge of all mankind under the terms of the natural law, provided the basis for regulation of the seas. In the first instance, the pope sought to lay out clearly zones of responsibility for the Portuguese and the Spanish in the newly discovered regions to block preemptively the wars in which they engaged during the 15th century from expanding into the New World. Within each zone, the Christian ruler would have the obligation of supporting the work of missionaries who would preach the Gospel to the inhabitants and, if necessary, civilizing them. In addition, should the inhabitants attempt to prevent the missionaries from preaching or peaceful merchants from entering and pursuing their interests, then a Christian ruler would have the right to use force to defend the travelers on the grounds that all men have the right to travel everywhere in peace. Furthermore, should the members of a society engage in behavior that violates natural law, the pope could order a Christian ruler to punish them and to bring that society's laws into conformity with the natural law.

In the 15th century several popes issued more than 100 bulls dealing with access to the Atlantic Ocean, the adjacent coast of Africa, and the islands discovered there.[45] Specifically, popes issued these bulls to regulate Portuguese and Castilian competition for possession of the islands that were being discovered and, potentially, for control of routes to the markets of Asia. These bulls and some commentaries on issues relating to this expansion that canon lawyers produced provided the theoretical basis for papal claims to regulate entry into the Ocean Sea and the activities of Latin Christians who went there.

When Pope Alexander VI (1492–1503) responded to a request from Ferdinand and Isabella that he legitimate Castilian jurisdiction over the islands that Christopher Columbus had recently discovered, he was operating within a longstanding legal tradition. The three bulls, *Inter caetera* (3 May 1493), *Eximiae Devotionis* (3 May 1493), and *Inter caetera* (4 May 1493) that he issued at this point contained the fullest expression of the body of papal and canonistic thought about access to the newly discovered lands that accompanied the process of discovery and expansion. According to the pope, the Spanish monarchs proposed "to bring under ... [their] sway the said countries and islands with their residents and inhabitants, and to bring them to the Catholic faith."[46] This placed the Spanish basis for entry into the Americas squarely within the traditional Catholic framework, namely, the fulfillment of Christ's Great Commission, Go teach all nations.[47]

The pope recognized that the Castilians, and presumably the Portuguese, had already expended a great deal of energy, wealth, and blood in advancing the interests of the Church, striving to ensure that "the Catholic faith and the Christian religion be exalted and everywhere increased and spread, that the health of souls be cared for and that barbarous nations be overthrown and brought to the faith itself." He alluded to the *reconquista* that saw the "recovery of the kingdom of Granada from the yoke of the Saracens" as a forerunner of what the Spanish intended to do elsewhere now that Spanish resources could be employed in the exploration of the newly discovered lands.[48] These sacrifices justified the pope granting the Spanish exclusive jurisdiction over access to the newly discovered lands.

Alexander VI placed two specific limits on the Spanish entry into the Americas. In the first place, the Spanish could not exercise jurisdiction over any lands that were already "in the actual temporal possession of any Christian king or prince . . . ." The second limit, formally created in the bull of 4 May, was the result of drawing a line from pole to pole "one hundred leagues towards the west and south, as is aforesaid, from any of the islands commonly known as the Azores and Cape Verde . . . ." The Spanish could operate to the west of the line while the Portuguese operated to the east.[49] The line was later adjusted by the terms of the Treaty of Tordesillas (1494), but the principle remained the same.

2.2.4. The world divided

Contrary to popular opinion, *Inter caetera* did not give possession of the Americas to the Portuguese and the Castilians. What the pope intended to do was to regulate the entry of European Christians into the New World in order to preclude conflict between the Spanish and the Portuguese, to regulate the entry of other Europeans into the New World in such a way as to ensure peace among the other Europeans who went there, and also to ensure orderly, peaceful contacts with the existing inhabitants of these regions.[50]

As long as there were only two nations involved in the exploration of the Atlantic world and as long as there seemed to be little of great value to be obtained in the Americas, the papal division of jurisdiction was successful because it was unchallenged. After the Portuguese began to acquire wealth from the Asian spice markets as a consequence of the voyage of Vasco da Gama in 1497 and after Cortes reached Mexico and its great treasure in 1520, however, the papal claim to determine who could enter the new worlds that were opening up to Europeans was not simply challenged but largely ignored.

In the first place, other Catholic rulers could argue that Alexander VI had exceeded his authority when he restricted access to the New World to Portugal and Castile. The King of France, Francis I (1515–1547), famously observed that "he would like very much to see Adam's will to learn how he divided up the world!" before accepting the papal claim to possess such a right.[51] The king's criticism of Alexander's bull was a part of the traditional Church-State conflict over the respective jurisdictions of popes and secular rulers and could have been made by any contemporary Catholic ruler.

2.2.5. Hugo Grotius and the protestant response

In addition to traditional Catholic criticism of papal pretensions to universal jurisdiction, the Protestant Reformation generated an even more serious objection. From the very beginning of the Reformation, the aim of the Reformers was to eliminate the papally led, hierarchically structured Church and return to what they saw as the less structured Church described in the Acts of the Apostles. During the Reformation the unified Church under the pope was replaced by a number of state churches each under royal direction. This led to the publication of numerous treatises on the right of a state or group of states to declare the Ocean Sea or part of it as closed.

The most famous and most influential of these treatises on access to the sea was that of Hugo Grotius. In 1608, he published anonymously a chapter of a proposed larger work as the *Mare Liberum*, a detailed refutation of the papal claim to regulate such access. To reject Spanish and Portuguese claims to their respective monopolies of access to the new worlds based on papal determination, he employed the arguments of the same writers that these Catholic rulers and their lawyers were using against them. He provided a point by point refutation of the claims contained in *Inter caetera*. The core of his position was that God willed that all mankind "to be of one race and to be known by one name" and that all men should "recognize their natural social bond and kinship."[52] Seen in this light, Grotius was arguing that rather than properly regulating good relations among men, *Inter caetera* encouraged division and, presumably conflict. Furthermore, Grotius pointed out that Alexander VI had acted as an arbitrator between Portugal and Castile so that his decision only affected them and no one else. Finally, he noted that "if the Pope has any power at all, he has it, as they say, in the spiritual realm only."[53] Grotius argued from within the framework of the medieval Church-State conflict so that even a Catholic ruler could accept his conclusions in good conscience.

Grotius's *Mare Liberum* and his *On the Law of War and Peace* formed one of the most important stages in the 16th-century construction of a body of international law based not on spiritual claims but on a rational analysis of the issues involved and enforced by the mutual self-interest of the major powers of Europe. Together with the Peace of Westphalia in 1648 that settled the religious wars of the 16th and 17th centuries without any reference to papal jurisdiction, Grotius's work marks the point at which papal claims to some kind of universal jurisdiction were no longer relevant to issues of legal order and regulation either within Europe or abroad.[54] Instead, the rulers of the major powers would meet and settle among themselves the conflicts that arose among them. *Inter caetera* and the entire papal-canonistic legal structure of international order was relegated to the dust bins of history.

2.2.6. The Papal world order in secular garb

What relevance then does the papal attempt to regulate access to the Ocean Sea have for the current problem of the regulation of space? What relevance can there be in claims that were rejected 5 centuries ago? One reason is that the issues that space travel raises are quite similar to those that Columbus's voyages presented to European rulers and there are various international agencies, especially the United Nations, that would like to play a leading role in the regulation of space. At the same time, as in the early modern world, there are governments whose officials would respond to UN assertions of regulatory jurisdiction in space with words rather like those of Francis I, and any modern government could produce scholars who could refute UN claims to universal jurisdiction in treatises like the *Mare Liberum*. The debate about access to the newly discovered lands that *Inter caetera* and the other texts that the discovery of the New World generated thus provide a kind of laboratory exercise in the course of developing international regulation of vast spaces.

Before concluding, however, let me provide an example of the way in which medieval notions of world order have already re-emerged in the contemporary world. In the 1960s a challenge to the Grotian position on access to and regulation of the sea appeared. Technological advances made possible the recovery of "polymetallic nodules" containing valuable ores from the bed of the sea.[55] This attracted the attention of the ambassador of Malta to the United Nations, Arvid Pardo, who gave a speech at the UN in 1967 "calling for the recognition of the area beyond the limits of national jurisdiction [over the sea] and its resources as the common heritage of mankind."[56] His theme was the "intolerable injustice of reserving the plurality of the world's resources for the exclusive benefit of a handful of nations" if the usual rules of international law, that is Grotian principles, were applied, that is placing the sea and its bed under the control of those countries that bordered it.[57]

Pardo proposed a resolution on the future control of the sea that would replace the traditional notion of jurisdiction over the sea with what he saw as a new one that reflected both economic interests and moral values.

> "The sea-bed and the ocean floor are a common heritage of mankind and should be used and exploited for peaceful purposes and for the exclusive benefit of mankind as a whole."[58]

In effect, the ambassador was suggesting formal recognition of the corporate nature of human society so that the sea's resources could be employed for the common good. Pardo's proposal would require reconsidering the meaning of state sovereignty and necessitate conceiving mankind as some kind of corporate whole under the jurisdiction of a supra-national authority. To a medieval historian Pardo's proposal suggests a secularized version of medieval papal and canonistic thinking about the nature of mankind and the possibility of a just world order.[59] The 'public order of the oceans' that Ambassador Pardo's proposal would overturn was the result of three centuries of legal thought and practice about world order.

2.2.7. Conclusion

In conclusion, let me state that there are two important points about access to the sea and access to space to discuss briefly: the claims to universal jurisdiction and the problem of enforcing such claims. The papacy claimed universal jurisdiction based on religious principles, the common origin of all mankind and a divine mission to preach salvation to all the descendants of Adam. The Grotian conception of order assumed an egalitarian state order regulated by reason and self-interest but limited to European states. In the 20th century, notions of universal human community and a universal regulatory jurisdiction have re-emerged in the form of the League of Nations and the United Nations. These agencies claim not a religious origin but a rational and humanitarian basis. While these agencies make no claim to convert peoples religiously, they have claimed the right, especially since World War II, to intervene in the internal affairs of societies on a humanitarian basis when internationally recognized humanitarian standards are being violated.[60] In effect, the current international situation sees the assertion of claims associated with the papacy, a reminder of the claim by some scholars that a good deal of modern political thought consists of secularized versions of medieval ecclesiastical thought.

The second point of comparison is obvious. Neither the early modern papacy nor contemporary secular agencies possess the power to enforce regulations that they issue. Enforcement always depends on the willingness of various states to see the own interests as supported by such enforcement. Like the regulation of the Ocean Sea, regulation of space will be rooted in and an extension of terrestrial politics. In the final analysis, it would seem that the regulation of space will fall to the most powerful states just as the regulation of the sea eventually fell to the great seaborne empires and earth-bound political realities will determine the regulation of space.

---

[39] Muldoon, James M. "Who Owns the Sea. Fictions of the Sea: Critical Perspectives on the Ocean". British Literature and Culture. Klein, Bernhard, ed. Aldershot: Ashgate, 2002.
[40] Ibid. pp. 22–25.
[41] Hanke, Lewis. The Spanish Struggle for Justice in the Conquest of America. Boston: Little Brown, 1949.

[42] Canning, Joseph. A History of Medieval Political Thought 300–1450. London: Routledge, 1996.

[43] Ullmann, Walter. "The Medieval Papal Court as an International Tribunal". Virginia Journal of International Law 11 (1971): 356–371.

[44] Muldoon, James M. Popes, Lawyers, and Infidels: The Church and the Non-Christian World, 1250–1550. Philadelphia: University of Pennsylvania Press, 1999. pp. 10–11.

[45] De Witte, Charles-Martial. Les bulles pontificales et l'expansion portugaise au XVᵉ siecle. Revue d'histoire ecclésiastique. 48 (1953): 683–718; 49 (1954): 438–461; 51 (1956): 413–453, 809–836; 53 (1958): 5–46, 443–471.

[46] Davenport, Frances G. and Paullin, Charles O. eds. European Treaties Bearing on the History of the United States and Its Dependencies to 1648. 4 vols. Washington, DC: Carnegie Institution of Washington, 1917. p. 62.

[47] Codignola, Luca. "The Holy See and the Conversion of the Indians in French and British North America, 1486–1760". America in European Consciousness, 1493–1750. Kupperman, Karen Ordahl, ed. Chapel Hill; London: The University of North Carolina Press for the Institute of Early American History and Culture, 1995. pp. 195–242.

[48] Davenport, Frances G., and Charles O. Paullin, eds. European Treaties Bearing on the History of the United States and Its Dependencies to 1648. 4 vols. Washington, DC: Carnegie Institution of Washington, 1917. p. 61.

[49] Ibid. pp. 77–78.

[50] O'Callaghan, Joseph F. A History of Medieval Spain. Ithaca: Cornell University Press, 1975. pp. 523–560.

[51] Morison, Samuel E. The European Discovery of America: The Northern Voyages, AD 500–1600. Vol. 1. New York: Oxford University Press, 1971. p. 435.

[52] Grotius, Hugo. "The Freedom of the Seas". Trans. Ralph van Deman Magoffin. Brown, James, ed. Scott. New York: Oxford University Press, 1916. pp. 1–2.

[53] Ibid. p. 16.

[54] Lesaffer, Randall, ed. Peace Treaties and International Law in European History: From the Late Middle Ages to World War I. Cambridge: Cambridge University Press, 2004. pp. 3–4.

[55] Mero, John. The Mineral Resources of the Sea. New York; Amsterdam: Elsevier, 1965.

[56] Brown, E. D. The International Law of the Sea. 2 vols., vol I, Introductory Manual. Aldershot : Dartmouth Publishing Co., 1994. p. 10.

[57] Pardo, Arvid. "Who Will Control the Seabed?" Foreign Affairs 47 (1968): 123–137.

[58] Ibid. p. 135.

[59] The notion that modern political thought is often rooted in secularized versions of earlier ecclesiastical thought is especially stressed in the work of Brian Tierney: see especially his Religion, Law, and the Growth of Constitutional Thought, 1150–1650. Cambridge: Cambridge University Press, 1982.

[60] For the current state of interest in humanitarian intervention: see the articles in the Journal of Military Ethics 5 (2006).

2.3 Celestial bodies: Lucy in the sky

Gísli Pálsson

2.3.1. Introduction

Ever since humans ventured beyond their cradle in East Africa, they have advanced their exploration of the rest of Earth step by step, achieving quite most spectacular success in the course of the last centuries. Now that most of the habitat of the globe has been charted and conquered, humans are turning their attention to the "remotest corners" of both living organisms and outer space. Focusing on the languages of voyaging and mapping, this article will discuss the similarities and differences of research on what biologists sometimes refer to as the "universe" within the human body (in particular the fetus and the genome), on the one hand, and the exploration of outer space, on the other hand. I will argue by zooming in and out that the endeavours of the big sciences of molecular biology and astrophysics are related in many ways – and not just by similar languages and metaphors.

2.3.2. Zooming in and out

In the last half a century, both molecular biology and space science have made immense progress. Some of the most significant events in molecular biology were Rosalind Franklin's photography of DNA (1952), the discovery of the double helix (1953), and the drafting of the human genome (2000). The most recent spectacular promise on this score is synthetic biology, aiming to design whole organisms practically from scratch. The description and broad implications of the intrusion of life into economics and politics is one of the most important issues on the academic agenda at the beginning of the 21st century. Among the momentous events that took place in space during the second half of the 20th century were the launching of Sputnik (1957), putting Gagarin in orbit (1961), landing humans on the moon (1969), and establishing the International Space Station (1998). To celebrate half a century of human exploration in space, *The Guardian* published a photograph (see Figure 1) with the accompanying text: "Astrophysicist Stephen Hawking, accompanied by his physicians and nurses, floats on a zero gravity jet at 24,000 ft above the Florida coastline. The plane made eight parabolic dips, giving

69

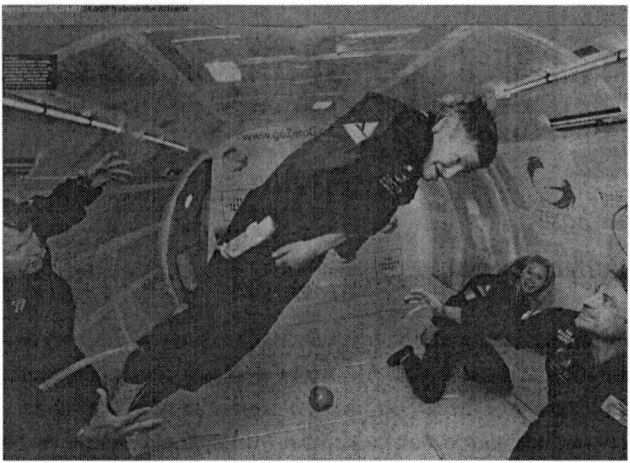

Fig. 1. *Astrophysicist Stephen Hawking, accompanied by his physicians and nurses, floats on a zero gravity jet at 24,000 ft above the Florida coastline. The plane made eight parabolic dips, giving Hawking the experience of weightlessness (source: The Guardian).*

Hawking the experience of weightlessness". The image powerfully illustrates both the mastery of space and the dependence and fragility of human beings. Newton's apple is not far off. Or is it perhaps a reference to the sinfulness of human digression into the heavens, a modern version of the story about the Tower of Babel?

Since the European Middle Ages, the notion of "celestial bodies" or "celestials" (from *caelum*, the sky) has usually referred to heavenly bodies belonging to the reigning gods or emperors on Earth. With the advances of molecular science, including those of genetic engineering and synthetic biology, the idea of con- structing corporeal bodies *for* outer space has become possible. While human exploration has hitherto taken place without subspeciation, this is unlikely to be the case in the event of human settlement in outer space. Indeed, the future may bring forth a post-human settler in space, a celestial "Lucy in the sky" – or, more likely, several kinds of them.

2.3.3. Fetal space

Several scholars have drawn attention to the parallels between the imagery relating to human fetuses and the womb, on the one hand, and, the celestial world, on the other. As Michaels points out, "Planets, supernovas, and galaxies have been showing up alongside fetuses, embryos, and blastocysts during the past twenty-five years, and their visualization occasions comparable journalistic indulgences and

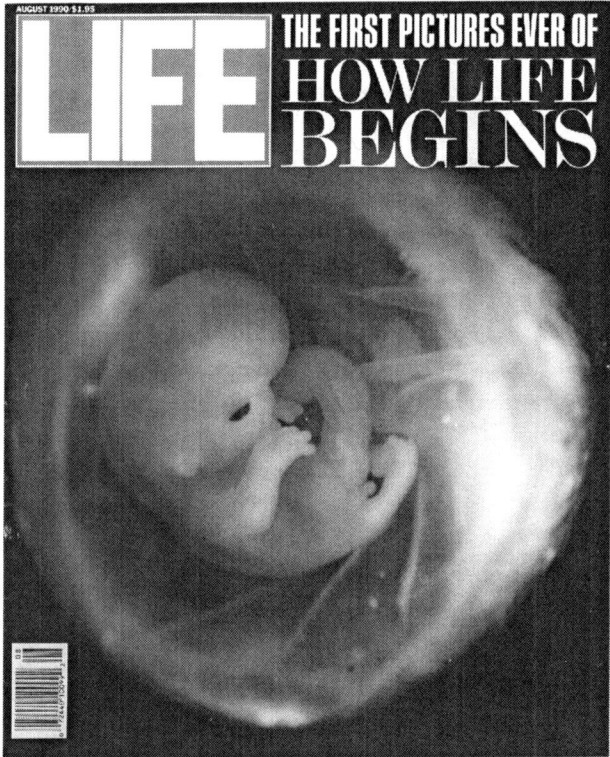

Fig. 2. *The first pictures of how life begins (source: Life Magazine, August 1990).*

epistemic quandaries."[61] Duden explores the path breaking photographs of Lennart Nilsson documenting "the beginning of life", published in *Life* magazine in 1965 and 1990, pointing out that the photographs, the first ones of the kind, and the accompanying text in *Life* repeatedly juxtapose fetuses and astronauts, bodies and space.[62] The 1990 issue (see Figure 2) elaborates on the first stages of human life by means of celestial language and imagery, starting with a fetus aged 2 hours:

> "Like an eerie planet floating through space, a woman's egg or ovum has been ejected by one of her ovaries into a fallopian tube... the luminous halo around the ovum is a cluster of nutrient cells feeding the hungry egg."[63]

Eight days later, *Life* goes on:

> "The blastocyst has landed! Like a lunar module, the embryo facilitates its landing on the uterus with leg-like structures composed of sugar molecules on the surface."[64]

The fascination with the fetus and outer space underlines human curiosity at the border of the unknown, the urge to extend sight by zooming in or out beyond the "natural" horizon of the human eye. New images and visual horizons, however, sometimes take on a life of their own, forming perception and politics. Thus, astronauts tend to speak of a gestalt shift as a result of their voyages into space, when seeing Earth from a distance, a "Gaia" perspective is created that seems to facilitate global, environmental concerns. No doubt, images from space, including images from Neil Armstrong's "giant leap for mankind" when landing on the Moon, have also formed public discussions on Earth in several respects. Likewise, the kind of fetal imagery presented by the famous *Life* photographs mentioned above probably had an enormous impact on the public discussions of biopolitics, in particular, on abortion and the rights of women. Duden addresses the issue with a reference to what she calls the "Nilsson Effect", stressing the shift in emphasis as a consequence of Nilsson's photography from the pregnant woman to the fetus and the resultant alteration in the power balance between those advocating "pro choice" and others in favor of "pro life".[65]

2.3.4. The universe within

A number of other works make similar analogies and connections between fetuses and astronauts, from the vantage point of outer space. An article by Sofia largely devoted to a commentary of the film *2001: A Space Odyssey* (1968) by Stanley Kubrick and Arthur C. Clarke suggests that the film establishes what she calls "Jupiter Space" through the imagery of the fetus, a space "whose contours are elaborated in visual complexes which equate the male brain, the womb, outer space, city landscapes, grids of light, microcircuits, the interiors of computers, skyscraper façades, and so on".[66] More generally, she argues that the human fetus serves as a symbol for Earth: "It is a cosmic symbol. It is not entirely inappropriate that the planet be represented by a signifier of unborn life, for it presently contains all of the possibilities for future life forms. From this perspective, disarmament might be seen as an act to prevent a cosmic abortion".[67] While Sofia's article is highly playful and speculative, it draws attention to military concerns and the Cold War and their impact for both biopolitics and space exploration.

With the development of digital technology, bioinformatics, and the new genetics, voyaging into the human body was escalated and extended to specific organs, in particular, the brain, cellular material, and the genome. Dumit concludes his ethnographic study of one kind of corporeal exploration, brain function imaging or "PET scans" (Positron Emission Tomography) which reportedly

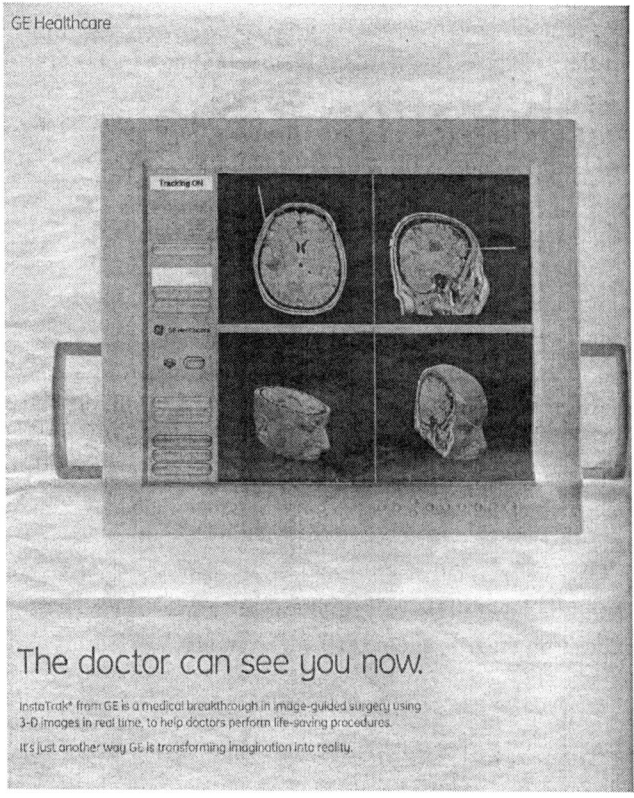

Fig. 3. *Advertisement "The doctor can see you now." (source: GE Healthcare).*

represent the human brain at work in different moods, states, and activities, with a similar note on the relevance of techno-scientific developments for the self-reflective anthropological project: "We...may have entered a space of active negotiation of the basic terms of our categories of the person. ...The use of these images in thinking about ourselves is in its infancy. *We* are at stake in this work. How can we not afford to risk jumping in and studying it?"[68] A recent advertisement for image-guided brain surgery says it all, "The doctor can see you now," presumably with the emphasis on "you" (see Figure 3).

2.3.5. Hunting and gathering genes

Eventually, "Man the Hunter" and "Woman the Gatherer", made in East Africa a long time ago, turned to hunting and gathering within the territory of their own

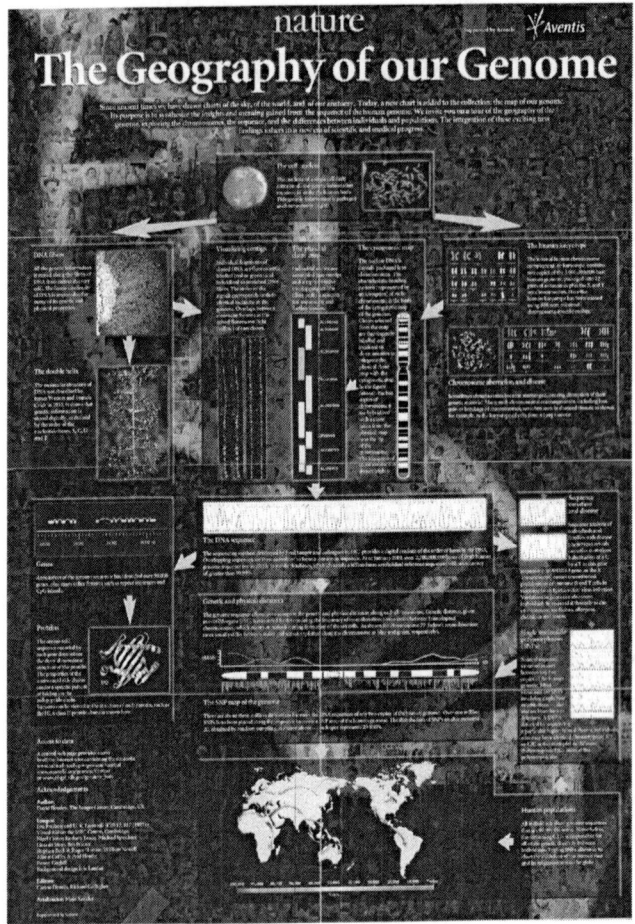

Fig. 4. *The Geography of the Human Genome (source: Nature).*

genome. What project could be more anthropological? *Homo viator* navigating waters right at home.[69] One of the characteristics of the genome era is the application of cartographic language to the human body. In year 2000, the first draft of a "map" of the human genome was announced. On that occasion, the journal *Nature* triumphantly published a poster, "The Geography of the Human Genome" (see Figure 4). *Nature* underlined its cartographic language by inviting its readers on a tour into the "universe within" with the following grand statement:

> "Since ancient times we have drawn charts of the sky, of the world, and of our anatomy. Today, a new chart is added to the collection: The map of our genome. Its purpose is to synthesize the insights and meaning gained

from the sequence of the human genome. We invite you on a tour of the geography of the genome, exploring the chromosomes, the sequence, and the differences between individuals and populations. The integration of these exciting new findings ushers in a new era of scientific and medical progress."[70]

Impressed with their ability to zoom in on the minute details and contours of hereditary material and their power of visualization, geneticists and molecular biologists have firmly reinforced their language of cartographies, a language that echoes the modernist notion of expansion and mastery. Textbooks represent a useful source of information on the language and imagery involved. At the turn of the new millennium, a standard textbook described molecular genetics with the grand terms of discovery and voyaging:

> "After many centuries, we have built up an approximate understanding of our external universe, but the universe within us has only very recently been the subject of serious study. The application of microscopy to the study of cells and subcellular structures provided one major route into this world, to be followed by pioneering advances in biochemistry and then molecular biology. Now, as we enter the next millennium, we are on the threshold of a truly momentous achievement that will have enormous implications for the future. For the first time, we will know our genetic endowment – the sequence of our DNA. Then our voyage into the *universe within* really will have begun."[71]

2.3.6. Elementary structures

With the expanding gaze of the new genetics, then, the hereditary signatures of individuals and populations, which are assumed to remain practically unchanged throughout the span of life, have become the subject of map making. Intensive research in the past decades has revealed what practitioners of the new genetics take as the elementary structures of the universe within and the geography of the genome. Each organism, in their language, possesses a genome that contains instructions necessary for constructing and maintaining a living example of that organism. Several kinds of maps are used in molecular genetics. It is conventional to distinguish between genetic mapping and physical mapping. Genetic mapping involves the application of genetic techniques to situate genes on chromosomes by relative distance ("genetic distance"). The first maps of this kind were constructed early in the last century by researchers working on the fruit fly, *Drosophila*, using genes as markers to establish the distinguishing features of the genetic landscape. Physical mapping, on the other hand, involves the use of molecular biology techniques to explore deoxyribonucleic acid (DNA) molecules directly, locating the positions of sequence features, including genes, on chromosomes by absolute distance (in units of DNA nucleotides). The mapping idea has often appeared in public discussions and the media.

The apparently innocent language of voyaging, of course, is somewhat misleading. Keller points out that while star-gazing has always been an important metaphor for biology it "certainly... has no place in the biology of today".[72] Gazing, she suggests, is increasingly enmeshed in actual touching, if not aggressive bombarding of the object in question. Perhaps, though, the development of cosmic maps beyond the Milky Way represents an appropriate parallel to genomic maps. Both projects have significantly altered the scale and meaning of the cartographic enterprise. In the case of cosmic maps, the subject of mapmaking is infinitely larger than the maps themselves, while for genomic maps the reverse is the case. One project puts parts into context, another zooms in on the parts.

2.3.7. Mappings

Maps, like photographs and other forms of visualization, are obviously powerful tools for scientists and other explorers of the unknown. Clearly, the mapping of genetic material has been greatly enhanced by the model of the double helix and the map of the human genome. At the same time, mapping deserves attention in its own right as the product of situated, historical activity. Maps have, indeed, become objects of critical attention in the social sciences and the humanities; giving way to a "spatial turn". Cosgrove attributes the extensive rethinking of maps to several factors, including the dissolution of a "euro-centric" geopolitics, changing techniques of seeing, and post-structuralist theorizing. For Cosgrove, the growing critical interest in maps and mapping corresponds to fundamental doubts about the grand narratives "and to the concomitant recognition that position and context are centrally and inescapably implicated in all constructions of knowledge".[73]

The notion of maps is used here in a fairly broad sense for any kind of visual surrogates of spatial relations. It is tempting to refer to genomic cartographies as "mere" metaphors and, indeed, an important avenue to explore in relation to the new genetics is: what do people mean when they speak of charting the "geography" or the "regions" of human genetic material? While this topographical language is clearly metaphoric in that it refers to the abstract visual representation of "crossover rates" (the layout of graphs and charts), it also seeks to realistically describe particular regions in the genome. In fact, a radical distinction between the "metaphorical" mapping of genomes and the "realistic" mapping of landscapes may not make much sense. The mapping of chromosomes – estimating their size, charting their layout, and positioning genes – is just as real a procedure as topographic mapping and, conversely, cartographic mapping may be just as metaphorical as genome mapping. Maps may look deceptively simple and straightforward. However, a critical dealing with maps is essential in the sciences and the humanities. Whether genomic, terrestrial, or celestial, maps are performative constructs, formed by the interests and perspectives of the people and the regimes involved in their making.

2.3.8. Out of Africa, out of Earth

In 1974, a 40% complete *Australopithecus afarensis* skeleton was discovered in the Awash Valley of Ethiopia's Afar Depression. The fossil (AL 288–1) was nicknamed Lucy after the Beatles song "Lucy in the Sky with Diamonds", which was played frequently at the time of discovery at the archaeological camp at Awash Valley. Lucy was estimated to have lived 3.2 million years ago, firmly establishing human origins in East Africa. For a long time, there has been a lively discussion of the roads out of Africa, their timing and significance for the understanding of hominid evolution and history. Recent archaeological findings from Dmanisi in the Republic of Georgia shed new light on some of the issues involved, possibly documenting the earliest members of the genus *Homo* outside the continent, thereby filling in significant gaps in current knowledge about a critical phase in human evolution.[74] Dated to 1.77 million years ago, the fossils involved indicate great variability in body and brain size, in the size range of *Homo habilis* and *Homo erectus*, reflecting, among other things, selection for improved terrestrial locomotion. These findings, then, provide an image of scattered and variable hominids adapting to time and place outside Africa. Will there be a parallel development in outer space, in the event of human settlement on different asteroids, space stations, moons, and planets?

One of the most important contributions to the anthropology of space is the volume *Interstellar Migration and the Human Experience* edited by Finney and Jones.[75] Written in the heat of the Cold War, under the threat of nuclear war, it represents a particular "galaxy of discourse", to borrow a term from Battaglia in another important contribution to the field.[76] For Finney and Jones, it is human "biocultural" nature to venture into new areas, to explore the entire globe and eventually head for the sky: "Sometime in the not too distant future, a space traveler will do something science fiction writers have been talking about for decades:

A human being will jump completely off a small world. Asteroids and small moons have very weak gravity indeed."[77] "Barring total nuclear war, a devastating collision with a comet or asteroid, or some other calamity on a worldwide scale," they conclude, "there is a good chance that this initiative will soon result in settlement in near space and that eventually our descendants will scatter among the stars."[78] No doubt, the birth of the first human child in outer space will be an event of enormous symbolic significance, underlining human *settlement* outside Earth in contrast to the relatively brief visits of the past, much like the birth of the first child outside of Africa and in the New World.[79]

2.3.9. Human evolution

The likely consequence of human settlement in space is one of the themes explored by Finney and Jones. For long, they suggest, the hominid species formed one interbreeding world population and speciation did not seemed possible. However, that would not hold in space since our descendants will probably be scattered throughout the vastness of space which would set up the conditions for the rapid speciation of *Homo sapiens*:

> "If the technology of space colonization really works, if our descendants do settle throughout the Solar System and then migrate to other star systems, humanity will never be the same again. The course of human evolution will change utterly and inalterably. ...The threshold of space is also the threshold to quantum biological evolution."[80]

Time, no doubt, has complemented and qualified the predictions of Finney and Jones on several scores. First, the obvious post-Cold-War potential event that they refer to as "some other calamity on a worldwide scale" is not a nuclear war but massive environmental change brought about by humans, including global warming. Our achievements at exploration, colonization, and resource use have made Earth a rather messed up place for humans and many other species, and unless the problems at home become paralyzing this is likely to escalate the exploration of outer space. Colonization and modernization have been taxing to the planet, a point emphasized by Mahatma Gandhi. Asked whether independent India would follow the British pattern of development, Gandhi replied: "It took Britain half the resources of the planet to achieve this prosperity. How many planets will a country like India require?"[81] How many planets will the world require to satisfy the needs of the masses and to clean up the mess?

Finney and Jones did not anticipate (at least not comment upon) the success of the new genetics and their potential relevance for settlement in space. Thanks to the spectacular advances of biotechnology and genetic engineering; long-term biological adaptation to outer space is no longer pure science fiction. As Rheinberger suggests, modern gene technology along with the molecular biology developed between 1940 and 1970 facilitate "the prospects of an intracellular representation of extracellular projects – the potential of 'rewriting' life."[82] The key tools of recombinant DNA work are not "sophisticated analytical and electronic machinery" but "macromolecules that work and perform in the wet environment of the cell. . . . The scissors and needles by which the genetic information gets tailored and spliced are enzymes. The carriers by which it is transported into the cells are nucleic acid macromolecules."[83] As a result, Rheinberger argues, the traditional dichotomy between "nature" and "culture" no longer makes much sense. Quite possibly, life will be "written" and "edited" for outer space in the future, inviting new kinds of citizenship and biosocialities, and new kinds of hybrids of technologies and organisms. Some of the early speculations along these lines are those made by Sofia in an article referred to above. If the Earth is an embryo, Sofia suggests,

" . . . then its womb is space. Although we know of no other living worlds, centuries of extraterrestrial fantasies capped by several decades of off-world practice have encouraged us to think of space as a good womb, full of inhabited planets. From this view, the Earth is just one of many cosmic pregnancies. It doesn't really matter if we abort it, for we can always escape to one of the new Star Children we pluck from the vacuum; we might even mutate into extraterrestrial cyborgs."[84]

2.3.10. The phenomenology of space

Not only does modern biotechnology increase the likelihood of human and post-human variation, encounters with radical "others" from space cannot be ruled out. Ufology, as Roth points out, the study of unidentified flying objects and extraterrestrial visitors, "is a discipline that has tried to understand racial diversity. Ufologists do not always call it 'human' diversity, but then the earliest European anthropologists were not sure that all speaking bipeds outside Europe were human either".[85] The difficulties of adapting to permanent settlement in space should not be underestimated, in particular, the damaging effects of radiation on human bodies and problems relating to what might be called the "phenomenology of space", the challenges posed by our Earth-bound perceptual, cognitive, and psychological capacities in the context of space. Addressing the relations between humans and the material world from a phenomenological perspective, Ingold suggests what he calls "a view from the open"; rather than imagining "that life is played out upon the surface of a world already furnished with objects", he argues, people "make their way *through* a world-in-formation rather than *across* its pre-formed surface. For that reason, the fluxes of the medium through which they move are all-important."[86]

Such a view emphasizes the every-day human experience of wind and weather. There is no compelling reason, however, to imagine that the "view from the open" does not apply to life in space; in fact, it may be even *more* relevant for astronauts than earthlings. Space is a particular kind of medium with its own formations and fluxes – a medium where everything floats, where solar particles ("winds") blur vision, and where Earth may not even be in sight. While space, much like the alien medium of water, poses particular problems for a species thoroughly adapted to terrestrial life, these can at least be partly eased by means of technological and computerized enhancements like robots and artificial intelligence. The mere fact that settlements in outer space *are* a possibility underlines that space is just as "natural" for humans as Earth.

2.3.11. Conclusion

There are profound problems to face in outer space, and, indeed, they may slow down attempts to establish human settlements outside Earth. Some of the problems likely to be encountered are unprecedented biopolitical and bioethical issues that will inevitably be distracting; issues relating to inequality, difference, citizenship, race, and, possibly, eugenics. Humans, however, have been moving extremely fast on the space front for the last half a century and their journeys seem likely to continue. Not only has curiosity been characteristic for the species from the beginning, leading to both the colonization of the entire planet and the exploration of the cosmos and the structures of cellular material, also there are good economic and environmental grounds for moving out of Earth. Given the spectacular advancements we have witnessed in recent decades in science and technology, including those represented by rocket technology, robotics, artificial intelligence, and the life sciences, human and post-human settlements in space are no longer simply the products of mythmaking and imagination. The original meaning of the title of "Lucy in the Sky with Diamonds" has always been a matter of debate, but whatever John Lennon's reasoning was when writing the lyrics – one theory states a psychedelic element induced by drugs (LSD) – Lucy's descendants seem to be heading for the sky.

[61] Michaels, Meredith. "Fetal Galaxies: Some Questions about What We See". Fetal Subjects, Feminist Positions. Morgan, Lynn M. and Michaels, Meredith, W. eds. Philadelphia: University of Pennsylvania Press, 1999. pp. 113–132.
[62] Duden, Barbara. The Nilsson Effect. Disembodying Women: Perspectives on Pregnancy and the Unborn. Cambridge: Harvard University Press, 1993. pp. 11–24.
[63] Ibid. p. 13.
[64] Ibid. p. 14.
[65] Ibid.
[66] Sofia, Zoe. "Exterminating Fetuses: Abortion, Disarmament, and the Sexo-semiotics of Extra-terrestrialism". Diacritics 14 (1984): 47–59.
[67] Ibid. pp. 48, 56.
[68] Dumit, Joseph. Picturing Personhood: Brain Scans and Biomedical Identity. Princeton: Princeton University Press, 2004. p. 185.
[69] Pálsson, Gísli. Anthropology and the New Genetics. Cambridge: Cambridge University Press, 2007.
[70] Nature. Oct. (2000): 407.
[71] Strachan, Tom and P. Read, Andrew. Molecular Genetics. 2nd ed. Oxford: BIOS Scientific Publishers, 1999. p. 295.
[72] Keller, Evelyn F. "The Biological Gaze". Future Natural: Nature, Science, Culture. Robertson, George et al., eds. London: Routledge, 1996: p. 108.
[73] Cosgrove, Denish. "Introduction: Mapping Meaning". Mappings. Cosgrove, Denis, ed. London: Reaction Books, 1999: p. 7.
[74] Lieberman, Daniel E. "Homing in on Early Homo". Nature 449 (2007): 291–292.
[75] Finney, Ben R. and Jones, Eric M. eds. Interstellar Migration and the Human Experience. Berkeley: University of California Press, 1985.
[76] Battaglia, Debbora, ed. E.T. Culture: Anthropology in Outerspaces. Durham, NC: Duke University Press, 2005. pp. 1–37.
[77] Finney, Ben R. and Jones, Eric M. eds. Interstellar Migration and the Human Experience. Berkeley: University of California Press, 1985. p. 5.
[78] Ibid. p. 333.
[79] On a sobering note, however, the saga of the Icelandic woman who gave birth to the first child of an Old-World couple in the New World some 500 years before the landing of Columbus has not received that much attention. This is the saga of Gudrid Thorbjarnardottir and her son Snorri Thorfinnsson, a saga that is documented and reconstructed in a recent book by; Brown, Nancy M. The Far Traveler: Voyages of a Viking Woman. New York: Harcourt, 2007.
[80] Finney, Ben R. and Jones, Eric M. eds. Interstellar Migration and the Human Experience. Berkeley: University of California Press, 1985. pp. 22–23.
[81] Moran, Emilio. People and Nature: An Introduction to Human Ecological Relations. Oxford: Blackwell, 2006. p. 150.
[82] Rheinberger, Hans-Jörg. "Beyond Nature and Culture: Modes of Reasoning in the Age of Molecular Biology and Medicine". Living and Working with the New Medical Technologies: Intersections of Inquiry. Lock, Margaret, Young, Allan, and Cambrosio, Alberto, eds. New York: Cambridge University Press, 2000. pp. 19–30.
[83] Ibid. pp. 19–30.
[84] Sofia, Zoe. "Exterminating Fetuses: Abortion, Disarmament, and the Sexo-semiotics of Extra-terrestrialism". Diacritics 14 (1984): 47–59.
[85] Roth, Christopher F. "Ufology as Anthropology: Race, Extraterrestrials, and the Occult". E.T. Culture: Anthropology in Outerspaces. Debbora, Battaglia, ed. Durham, NC: Duke University Press, 2005. pp. 39–93.
[86] Ingold, Tim. "Earth, sky, wind, and weather". Journal of the Royal Anthropological Institute 13 (2007): 19–38.

2.4 Why we had better drop analogies when discussing the role of humans in space

Sven Grahn

2.4.1. Analogies used in the early space age to define the role of humans in space

Humans always try to use analogies when new technologies appear. For example, in computer usage we talk about mail, files, libraries, folders and other terms that we recognize from everyday life.

Space flight, when it was introduced 50 years ago was a completely new realm of human activity. To understand how this new field of human activity should be incorporated into society various analogies were used. Some analogies seem to hold and others do not. But anyhow, analogies are powerful tools in a situation like this. Here are some of the most prominent analogies used to describe human space flight in the early era of space travel.

1. Explorers of New Worlds (America), Settlers
2. Sailors on the New Ocean
3. Pilots of spacecraft for reconnaissance, intercept, strike[87]
4. Soldiers on a battleship
5. Scientists on a field trip or in the laboratory
6. Factory workers

2.4.2. Explorers of new worlds

This metaphor is very powerful and appealing to the human spirit. Early space travelers were compared to Columbus and his expeditions to the Americas. It is fascinating how this analogy still captures people's imagination despite the fact that the exploration of the Americas was done with a direct profit motive, while human space travel obviously is far from bringing home such immediate rewards. To support this analogy, the extraction of Helium-3 for use in, as of yet, non-existing fusion reactors on Earth is often mentioned. In the 1970s, the idea of building large solar-power stations on the Moon and sending them down to geostationary orbit was used as one rationale for exploring the Moon. In this period, the notion of artificial human settlements in space (at the L5 point) was also advanced by Dr Gerard O'Neill in 1974. In this period books, such as "Limits to Growth" published by the Club of Rome", very much influenced many thinkers and clearly played an important role behind O'Neill's formulation of his grandiose space settlement concept.

2.4.2.1. Sailors, pilots, and soldiers

" . . . the situation was aggravated by the discrepancy between the terrestrial stereotype of relative movement and the reality of space flight: we got used to rely on our experience of operating airplanes and automobiles, where it is possible to "add gas" to catch up with a moving object. . . . Besides, a significant role in guiding an air plane belongs to intuition. . . . But I am not sure that space guidance could rely on intuition. In order to predict relative movement of objects, it is necessary to know their orbits precisely; it is impossible to rely on anticipation . . . "

Analogies (2–4) [Sailors on the New Ocean, pilots of spacecraft for reconnaissance, intercept, strike, and soldiers on a battleship] are somewhat weak since there is no freedom of movement in space such as on the ground, in the air or at sea. You cannot change your destination or course in space at will. Orbits are rather "rigid" paths. Let me quote from a book by a colleague of Valentina Tereshkova's, Valentina Ponomareva in, "The Female face of the Cosmos":

"Also, the military role of humans assumes that there are valid military targets in space that could be engaged with manned spacecraft. If there were nuclear weapons or manned reconnaissance bases or "battle stations" in space, there might possibly be a role for spacecraft for reconnaissance, intercept and strike. However, the huge number of nuclear weapons that could conceivably be deployed in orbit makes it well nigh impossible to engage them all with interceptor, manned or unmanned. Inspecting and destroying reconnaissance spacecraft could be done with a manned space interceptor/strike vehicle, but the relative value of having humans in the loop is debatable."

2.4.2.2. Dropping bombs from manned battle stations

Manned battle stations equipped with nuclear weapons was even proposed by Sergei Korolev (the "father of Sputnik") in the early 1960s as a means of offsetting the short flight time to the U.S.S.R from U.S. nuclear missiles based in Europe. Soviet battle stations with nuclear warheads that could be dropped on the U.S. would permit a short delivery time to targets in the U.S. Of course, several orbital battle stations would be needed, and the opportunity to drop the bombs would not be available at all times. But, once the bomb was dropped from orbit, it could, theoretically, reach targets within minutes. However, this concept was short-lived and very much influenced by events at the time. Of course, Soviet leaders sought a much more direct approach to solving the problem – they put missiles with nuclear warheads in Cuba . . . and the rest is history!

2.4.2.3. Manual vs. automatic control

> " . . . the leadership of the Missile Forces has more trust in automatic satellites, and it underestimates the role of human beings in space research. It is a shame that in our country, which was the first to sent man into outer space, for four years the question has been debated whether man is needed on board a military spacecraft. In America this question has been resolved firmly and conclusively in favor of man. In this country, many still argue for automata . . ."
> (Translation by Slava Gerovich)

The limited value of offensive military action in space was quite clear rather early in the space age, but these notions still drove the debate about how much the space pilot should intervene in the maneuvering of his space vehicle. It is interesting to quote from the letter to Party leader Leonid Brezhnev that Yuri Gagarin and some famous cosmonaut colleagues wrote in October 1965. They saw the military role of men in space as important and concluded that the U.S. had already decided that astronauts had military tasks in space;

In the Soviet Union, the role of man in controlling spacecraft as pilots was an issue for debate. The first piloted spacecraft, the Vostok and the Voskhod, were almost entirely automatic with back-up manual control systems. As the space exploration agenda became more ambitious and included such complex tasks as rendezvous, docking, and lunar landing, designers faced the problem of optimal division of functions between human and machine on board. While the U.S. space programme gave the astronauts primary responsibility for these tasks, the Soviet designers largely continued their reliance on automatic systems. They believed that the reliability and functionality of piloted spacecraft were largely dependent on the technical characteristics of automatic systems, and they saw the automation of spacecraft control as a complete replacement of human activity with automatic devices. The emphasis on automation in the Soviet Union has been ascribed to the fact that spacecraft were developed by "artillery men" and not by aircraft designers. The latter would have put emphasis on manual control in analogy with piloting aircraft while "artillerymen" had no such prejudice . . . In the U.S. spacecraft were developed by the organizations and companies deeply rooted in aeronautics.

Thus, even though only obliquely related to the functional role of man in space, the issue of automatic or manual control is interesting to examine also in a political context. As far as I can remember there was an undertone in reporting about manned space flights in the 1960s. The strong reliance of manual control by the U.S. was portrayed as a political statement – free man navigating freely on the New Ocean, in contrast to the passive Soviet cosmonaut carried along by the spiritless automaton of an oppressive society. Little did we know that the difference in approach may have had to do with the different technical approach between two branches of the military industry; artillery and aviation.

2.4.2.4. The human spy-in-the-sky

The manned reconnaissance satellite, an analogy to the observation balloon and the scouting plane with a photographer in the back seat, had long life. In the U.S., the Manned Orbiting Laboratory (MOL), and Air Force project to put a manned spy satellite into orbit replaced the manned boost-glide prototype space plane Dyna-Soar as the primary manned military space project in the U.S. However, the rapid advance of unmanned reconnaissance satellites and the advent of digital cameras made manned spy satellites less attractive and MOL was cancelled in 1968. However, in the Soviet Union this idea was pursued vigorously and the Almaz experimental military space station was flown twice during the 1970s to evaluate the usefulness of "man-in-the-spy-loop".

In Almaz, man had the role of a "relevance filter". Instead of taking numerous, possibly uninteresting pictures of wide swaths of ground, the cosmonauts would take pictures of targets that they could see themselves through the viewfinder and they would develop the film on board and examine the pictures to see which ones were worth sending to earth – a reasonable role for man, one could say. There were two means of sending the valuable pictures to the ground; a TV link to the ground with scanned images, and a capsule that would drop the actual film to the ground. The 20-ton Almaz complex was grandly designed, but even it proved too cumbersome and lacked cost-effectiveness compared to automatic spy satellites that continued to rely on the photographic technique. Exit man as a spy-in-the-sky.

2.4.3. Man as attendant at a staging point in space

However, the metaphor of space as a new arena for travel was a strong driver for early space station work, even for ISS. Staging points, like stage-coach stations where man and beast could recuperate and obtain supplies, were even discussed. Remember the "Space Operations Center", Boeing's concept from the 1980s? It was envisaged as a fuel depot and parking garage for space tugs that would take Shuttle-launched payloads to destinations like geostationary orbit or the Moon. As it turned out, this concept faltered due to the same circumstances that the Space Shuttle became of a victim of – the cost of unmanned expendable vehicles is lower than man-rated craft and equally reliable. The manned launching system has not proven considerably more reliable than the expendable launch vehicle – it seems that technology in this field has stagnated for 40 years . . . or has it?

2.4.4. The scientist in the field or in the laboratory

The scientist on a field trip is still a strong metaphor that drives our thinking about what to do next in human space exploration. The scientist in the laboratory is, of course, the main theme for activities on the Internal Space Station (ISS), even though the astronauts are mostly kept busy with maintenance tasks. The first serious scientific tasks carried out by space travelers actually took place on the last Apollo flight to the Moon in 1972, when the first geologist, Dr. Harrison Schmitt, got to make the trip. The Skylab space station in 1973–1974 was also a very ambitious attempt to do astronomy and earth observation from orbit with space travelers as key actors. Perhaps this one of the few analogies that holds.

2.4.5. The factory worker

This was strong metaphor in the 1970s and 1980s when the microgravity factory was a major selling point for both the Space Shuttle and the ISS. Undoubtedly, important scientific results have been achieved in microgravity, but no large-scale manufacture of products unique to the space environment has taken place.

2.4.6. Modern analogies for the role of humans in space

As we entered the 1980s new analogies were used to describe what was perceived as the key roles of humans in space. These analogies were not as bold as those of the early space age

1. The maintenance worker (trouble-shooter)
2. The construction worker

2.4.6.1. "Cosmic truck driver" and "trouble-shooter"

Clearly, humans are unbeatable as trouble-shooters or maintenance workers on complex space infrastructures such as Mir, ISS and the Hubble Space Telescope. Such huge investments are hard to maintain remotely, they need physical intervention. The Hubble space telescope, an immensely valuable piece of hardware, is a fascinating case. It was designed to be "man-tended" and use the basic character of the Space Shuttle. The Space Shuttle, as designed, embodied the notion of the astronaut as a cosmic truck-driver and tow-truck operator. The re-usability of the Space Shuttle would provide low-cost space transportation and the human intervention would ensure successful deployment and repair of valuable space assets. Dramatic salvage operations in orbit were performed during the first 15 years of Shuttle operation, but reality finally caught up with the unrealistic estimates of Shuttle launch costs.

What remains of this vision of the role of humans in space as the trouble-shooters. Hubble is a convincing example of this and the assembly of ISS shows what people can do when automatic systems fail. What remains to be discussed is, of course, what is the future of huge future space infrastructures that will need this kind of human servicing and maintenance?

2.4.6.2. What happened to telepresence?

In this context it is worth remembering a catch-phrase from the 1980s – "telepresence", i.e., remotely controlled manipulators with tactile abilities.

With "virtual reality goggles" earth-bound human operators could perform tasks that astronauts do. There could be extra crew-members in the shape of telepresence robots. What happened to "telepresence"? Have you heard of it lately? Hasn't technology advanced so much that it is now much more feasible?

Let me be clear; as I understand it "telepresence" is not pure robotics; it is the extension of a human presence, a much more capable agent than a robot. Because, as Robert Zubrin[88] has so eloquently said:

> " ... *Two hundred years after Lewis and Clark, there is not a robot on this planet that you can send to the grocery store to pick up a bag of unbruised apples. If they can't do a trip to the grocery store, how can they explore a planet? How can robots match the intuition, versatility, ingenuity, and common sense of the human explorer? ...* "

2.4.7. How is the role of humans affected by the cost/risk aspect of space flight?

Clearly, with the immense costs of putting people into space, the work that is entrusted to humans must be high pay-off tasks that can only they can perform. Complex assembly and trouble-shooting clearly falls into this category. But what would happen if the transportation costs could be reduced by a factor of, say, ten or a hundred? The mass penalty, in absolute terms, of carrying people into space would be reduced, but the relative mass penalty between carrying people or automatic devices would be the same as before. Nonetheless, a lower transportation cost might open up the area of space salvage and repair to more human participation. Space tourism where no "useful" tasks are performed is an obvious field that could emerge, but the risk of getting killed must be reduced to levels comparable with that of extreme sports like parachute jumping. But what are the chances of such radical changes in the basics of space transportation? Slim I'm afraid.

2.4.8. Conclusion

Let me, as a concluding remark, quote a fellow space cadet David Portree who argues that we should choose goals for human spaceflight that are relevant to people everywhere. In a recent posting on the wonderful Internet Forum Friends and Partners in Space (FPSPACE), he proposed:

"Discovering life elsewhere, protecting life on Earth – those are relevant goals."

Let us drop the analogies and concentrate on Mr. Portree's goals!

---

[87] Gerovitch, Slava. "Human-Machine Issues in the Soviet Space Program". Conference on Critical Issues in the History of Spaceflight. National Aeronautics and Space Administration (NASA) Office of External Relations, History Division (NASA SP-2006-4702). Washington, D.C. 2006.
[88] Zubrin, Robert. "The Human Explorer". The New Atlantis 4 (2004): 93–96.

CHAPTER 3

"SPATIALITY" – SPACE AS A SOURCE OF INSPIRATION

3.1 Summary

Olivier Francis

Human space exploration has been always part of the dreams of human beings. This is a challenging adventure that could not be conducted by one nation only. It will require a planetary effort and an international cooperation. This part is dealing with the space as a source of inspiration.

The first presentation offers a critical and thorough analysis on how we talk and write about space. Ulrike Landfester demonstrates through the analysis of a short text from NASA about their mission that it (in her own words) "carries implicit historical meanings which gives the Unknown an anthropomorphous shape". She is convincing in claiming that the way we talk and write about space is somehow fictional (we need to be aware of it) and that Humans in space will certainly meet out there someone completely different from us in all the aspects (knowledge, learning, body and spirit). She points out the dangers and the remedies.

The second paper by Nicolas Peter reviews the different phases of the space exploration. Following the increase of space agencies, space exploration is no more restricted to a few countries. The ranges of space capabilities are varying with the countries. It results in an increase of long-term collaborations between nations and an international coordination of global space exploration. The motives are foreign policies and technology developments. It also aims to acquire knowledge in hard sciences as well as in humanities and social sciences.

A few recommendations could be drawn from this part. First, we should not neglect the key role of the humanities and social sciences in space exploration. Secondly, this new adventure will be possible only through an international cooperation including all the active nations in space as well as through a challenging and inspired international coordination between all the actors.

Finally, our fascination for humans in space comes from our endless curiosity about our environment. I found this session very instructive; I hope you will feel the same. Ultimately, space exploration is our only way out from our dying solar system, isn't it?

3.2 Missing the impossible: how we talk and write about space

Ulrike Landfester

3.2.1. Introduction

"It's late at night, and you receive an urgent phone call from the White House. 'The President wants to know why we should continue to put humans in space. He wants a one-page summary on his desk by tomorrow morning.' What do you write?"[89] These are the opening words of Michael Huang's article on *The Top Three Reasons for Humans in Space* which appeared in *The Space Review* in April 2005. Huang, who runs the website *Spaceflight or Extinction* – its title is based on a quotation from Carl Sagan's book *Pale Blue Dot: A Vision of the Human Future in Space* (1994)[90] – may certainly be assumed to be biased strongly in favour of sending humans to space. This initial question, however, whether intentionally or not, puts a precisely pointed finger on one of the more problematic issues of humanity's urge towards expanding their existence beyond the limits of the planet Earth: when political and financial stakeholders of spaceflight are to be persuaded to engage in sending humans to space, they must be presented not only with a meaningful and transmittable vision but also with arguments which convey and stabilize its practical viability.

This paper is going to propose the following argument: as the medium in which both vision and arguments are developed is usually that of written or spoken language, the product to be presented to its addressee or addressees – in this case the U.S. president – is structured not only by the facts which it conveys content but also inevitably by the additional meanings inherent in the terms used. These meanings are evoked and enforced by the terms' historically grown semantics that stem from contexts other than the specific issue in question and unavoidably influence the issue. Following Marshal McLuhan's famous dictum "The medium is the message",[91] this connotational influx of meanings is a non-negotiable key element of the medium – language – and as such crucial to the message 'humans in space'. It is, to put it pointedly, impossible to create discourse without this dialectic relationship between medium and message. This is especially true in the case of a message concerned with something about which we know next to nothing, i.e., the implications and consequences of humans not only travelling for a short time but also actually existing, living, working, reproducing in space. Here, this particular structural impossibility is reinforced by the discourse's as of yet uncharted horizon of reference. Thus, Huang's question "What do you write?" is a short version of a whole complex of questions, asking whether the writer knows precisely what he or she is doing when encoding space in language, when choosing specific words instead of others, using metaphors or, even more important, conceptual terms which are colloquially comfortable and seemingly self-evident in their meaning. However, these terms, in fact carry semantic weight that cannot but cross the borders of the terms' pragmatic situational content, and in turn influence this content from beyond these borders, placing it in a context which may enhance or subvert, but will certainly invest it with more than the terms' literal meaning.

3.2.2. The Motto's Mission: a case study

Let us take, for example, the motto which can be found on the NASA homepage under the heading 'Mission': "The more we know about the universe, the more we learn about ourselves. From satellites monitoring our planet's resources to orbiting observatories monitoring deep space, every NASA mission embodies the spirit of discovery."[92] This short text carries the semantic weight of at least five concepts which have, from the purely ontological point of view, just about nothing to do with the object the motto pertains to – the universe – and everything with the way historically grown meanings have come to shape our perception of space: 'Knowledge', 'Learning', 'Ourselves', 'Embodiment' and 'Spirit'.

An analysis of the historical development of these contexts and of the use they are put to by the NASA mission statement shows that reflecting on the ways we talk and write about space by means of the heuristic instruments used and honed by the Humanities can make an important contribution to deal with the challenges presented by the exploration of space. Knowing precisely what we do when we talk and write about space will enable us to recognize, and consequently, to overcome the limits of our perception of the Unknown Other. In this manner, crossing the borders into space will not simply repeat the colonization patterns inherent to our traditional 'poetics of discovery' as, for instance, implemented by the developments after the Great Encounter of 1492 but rather makes it possible and indeed challenges us to discover alterity as something to be aware of, to respect and to adjust to productively.

3.2.2.1. Knowledge

The concept of knowledge we automatically associate in a context like that of the NASA motto is one that is or rather understands itself as being based on solid scientific evidence, meaning that what we know is something that exists independently outside our subjective perceptions. Recent debates, however, have readjusted the ontological notion of knowledge as something that is already there and only needs to be retrieved, towards the epistemological notion of knowledge as something that is produced by sets of practices, mechanisms, and principles, assembled by structural affinities, necessity, and historical coincidence and controlled by strategies of perception inherent to their approach to their specific goal. In short, if we talk about knowledge we talk about ways to talk and write about space, conventionalized nowadays in the name of scientific evidence – but this convention in itself is a historically grown one, which has to be revalidated time and again.[93]

To illustrate this it is helpful to look at the way the biblical writings of the Old Testament, to be precise: the Genesis narrative deals with both terrestrial and extraterrestrial space, as the knowledge generated by the Bible in this area as in many other respects shaped scientific thought up to the 18th century when finally the notion of the Bible transmitting the true voice of God was replaced by the realisation that it was written by humans. This realisation was instigated and promoted by the then emerging Humanities in the shape of philological and historiographical approaches to the text of the Bible which proved that the inconsistencies of this text were due not to God's playfully enigmatic encoding of esoteric truths, but to the human factor – errors as well as creative imagination – in narrating, writing and translating early Judeo-Christian history.

There are two significant passages on space in the Genesis narrative. The first is the story of Noah's ark as related in Gen. 6–8. Having expelled Adam and Eve from Paradise and decreed the hardship of labour – labour of the land for Adam, labour of childbirth for Eve – as a suitable penalty for having eaten from the tree of Knowledge, after some time God finds that the tribes that sprung from Adam and Eve have started to enjoy themselves with the promiscuous taking of lovers, ignoring the strictures placed upon them, and he decides to exhaustively cleanse the space of his creation from these degenerates. Only Noah is spared and ordered to build the arc which then becomes the vessel in which God's living creation – animals and humans – survive. This vessel, as a closed-in space wholly dependent on God's Will, is a symbol for the radically monologue type of knowledge God wishes to instil in the survivors. It is the knowledge that the whole of the space created by God's Word is subject to his authority exclusively, not to be challenged by human perception or intelligence, and this is the paradigm that dominated the sciences until early modernity: whatever the scientifically validated evidence concerning natural phenomena, including them into the space of biblically generated knowledge made it imperative to bring them into line with God's prerogative of giving meaning to creation, this stricture being so forceful that Isaac

Newton even in 1706 put forth the idea in his work *Optics* that the concept of 'space' must be thought of as God's sensorium for the relations between all parts of his creation – even if, in the later editions, he inserted an 'as if' into the relevant passage to put some distance between the dogmatic belief in God's omnipotence and his own scientific findings.[94]

The second narrative referred to, also in Genesis, tells the story of the tower of Babel. This time, the threat to God's order of space is inspired not by humankind's frivolous zest for living, but by the desire of Noah's descendants to create a symbol for themselves that might represent the unity of mankind: "They said, 'Come, let's build ourselves a city, and a tower whose top reaches to the sky, and let's make ourselves a name, lest we be scattered abroad on the surface of the whole earth.'"[95] Yahwe, it is recorded, then "came down to see the city and the tower, which the children of men built" and was not amused: "Behold, they are one people, and they have all one language, and this is what they begin to do. Now nothing will be withheld from them, which they intend to do."[96] The end of the story is well known – God "confused the language of all earth"[97], so the builders should not understand each other and stop reaching out into a space, which was not for them to reach out into, it being the privileged space where God Himself dwelt. Of course this story first and foremost served the purpose of rationalizing the historical fact that there were different languages spoken among peoples which the Old Testament claimed to be all of the same origin, but it is nonetheless significant that this rationalization should be couched in an image which binds space to language and vice versa, implying that had humankind retained the unity of language it might have long since reached the sky and become humans in space.

Fig. 1. *Pieter Brueghel the Elder, The Tower of Babel, 1563, Kunsthistorisches Museum, Vienna.*

What these two paradigms shaping the meaning of humans in space have meant for the history of science is most clearly illustrated by an effort made by the Jesuit priest Athanasius Kircher to bring scientific evidence to bear on God's seemingly wanton destruction of human achievement. One of the key figures in the process which on the threshold to modernity strove to reconcile God's authority with the emergence of autonomous scientific thought, Kircher in 1679 in his book *Turris Babel* did the maths on the Babel tower project and came to the following conclusion: as the distance between earth and sky was, in his reckoning, 265,380 km, 4.5 million men would have had to put together 400 trillion of bricks over a time span of 3,400 years to finish the tower – and by then there would not only have been considerable problems of engineering but, much more important, the weight of the tower would have forced the earth from its place in the middle of the universe, which in turn would have ultimately destroyed it – so God was fully justified in stopping the builders, as he only wanted to save the world.[98]

It was only after Galilei and Kopernikus between proved that the earth and with it man was not at the centre of the universe and thus triggered the most memorable identity crisis of pre-secular Western humankind that knowledge, in the course of the 18th century, started to become something to be gathered and evaluated unhampered by metaphysical determinism – or at least something that defined itself as purely scientific in the sense of the word 'knowledge' as used by NASA. To serve this purpose, the concept of knowledge had to shed its connections with what had once been its wellspring, i.e., theology, the Arts and the Humanities: by the end of the 19th century scientific knowledge had been purified at the cost of losing sight of the fact that the very medium within and through which it existed, namely language, however pure the scientific interest, retained, and still retains today, the potential of historically grown polyvalence, especially when crystallized into conceptual terms like 'knowledge'.

3.2.2.2. Learning

Since the Middle Ages, learning has first had the meaning of acquiring knowledge of the Holy Script so as to be able to affirm its content. Then, as already mentioned above, approaching pre-modernity it meant reconciling scientific evidence with the dominant framework of Christian dogma. It was only after the aforementioned identity crisis that a concept of learning began to emerge during the 18th century, which, coupled with the ascent of reason as the leading category of humanness, became a technology of self-modelling. This concept is an offspring of European Enlightenment, when man finally left the idea of himself as a puppet of God behind and took over responsibility for the shaping of both his own history and history as such. However, learning then, in the early modern times, always encompassed the acquisition of an awareness of the rhetoricity of all knowledge: public reasoning as promoted by Kant in his famous essay on Enlightenment[99] meant not only to be sure of facts but also to be able to configure them according to a logic of discourse which was conscious of itself, of the fact that talking and writing were bound to historical patterns of perception which functioned more like images than like, for example, mathematical diagrams.

Not long ago, Dan Brown's novel *The Da Vinci Code* (2003) gave a concise resumé of how learning in the Western world was and still is bound to what may be called the Image Factor. Centred around the Louvre in Paris as a space where not only the beauty of culture but also the knowledge about the construction of history by means of such beauty is stored in thousands of masterpieces, the narrative tells the story of the deciphering of the da Vinci code as a story of learning how to look beyond the merely beautiful towards the patterns of thought embedded therein. What is of interest here is a subtext not immediately evident: in Leonardo da Vinci, Brown chose an artist who, like Kircher, stood at the threshold between pre-modern affirmative and modern critical learning, and who thus allowed Brown to

Fig. 2. *The Apple Macintosh logo.*

show how closely art and science were and still are related to each other. It is not by mere chance that the process of learning the protagonists go through during the narrative begins with an artist who was one of the first to construct flying machines – like, for example, a helicopter as documented by several drawings – and reaches its climax with Newton's apple which allows the protagonists to finally break the cryptex code, Newton being the one to first define space as a physical reality. And it is equally no mere chance that Apple Macintosh uses the fruit from the biblical Tree of Knowledge as its logo, bitten into, of course, as the users of Apple Macintosh computers are thus implicitly advertised as having acquired superior knowledge avant la lettre by simply buying – according to its promotion – superior hardware.

This connection, which may or may not be a subtle way of product placement in Brown's novel, anyway underlines once more the point Brown is trying to make, i.e., that learning means nothing like an acquisition of fixed meanings but rather the ability to take their hidden agenda into account, weigh them against each other and reach the conclusion that, preposterous as Brown's theory of Jesus having spawned a genealogical line which is still extant may seem as such, in the age of spaceflight, which was anticipated by da Vinci and prepared for by Newton, there may be a lot more to be learned from the arts than meets the eye. Precisely because their curiosity in the 'real world' – and, intrinsically connected with this, the conditions of what may be termed 'reality' and why – stems from the same wellspring as that of the natural sciences.

3.2.2.3. Ourselves

The notion of man as an indivisible entity, an identity or self, is a product of the Enlightenment as well and can be found reflected in the early modern notions of space: it was René Descartes who first developed the idea of two different, but coexisting concepts of space, one which manifests itself in the material world outside the thinking individual – *res extensa* – and one which exists as the inner space of thought, the *res cogitans*, the realm of the thinking individual's perception of space. After Newton had scientifically defined this outer space, the Cartesian idea of the two spaces became the origin of what until well into the 20th century remained a bone of contention in the debates on space, as this idea opened up the question of whether man in thought realizes what is ontologically outside him, space that "is", or whether man constructed space according to topological premises, so that space even in its physically evident aspects always remains bound to man's topological perception, an effect ratfher than the cause of the notion of space and thus a projection of ourselves.

Stanislav Lem in his 1968 novel *Solaris* quite decisively went for the second option. The astronaut Kris Kelvin is sent to the planet Solaris to join the crew stationed there and finds his colleagues housing guests who, as he begins to realise when these guests are joined by Kelvin's dead wife, are generated by the ocean outside the station which is obviously a living thing in its own right and with its own inexplicable powers. Lem's novel leaves open whether this ocean actively reproduces shapes stored in the crew members' memories by telepathic communication with the human minds or whether the fact that they seem to recognize these figures is an effect of their emotional projections on creatures which serve as a kind of imagination-activating screens. The moral of the plot is that, in Lem's opinion, humans in space need to be aware that they tend to meet the Unknown Other by projecting 'ourselves' onto or into it, thereby creating potentially dangerous, even lethal situations for themselves, because they are not equipped to recognize that this Unknown Other may well be nothing like 'ourselves', that it may have neither identity nor sociality, not even something like what we think of as intelligence; it may neither wish to communicate nor lend itself to cooperation, it may not even be alive in the sense of the term as we use it.

3.2.2.4. Embodiment

Giving a body to something, which has none yet makes it visible, defines its location in space. This in itself is nothing new to modernity, as already Aristotle had argued for a concept of space consisting of bodies in relation to each other – but where Aristotle had used the term 'body' in a purely physical way, meaning the material extension of any given thing or substance, today the term 'body' is colloquially linked to the human body – also an effect of the 18th century reshaping the idea of humanity, when the human body began to be recognized – and one might argue, constructed – as existing in a position where culture and nature, individuality and sociality interact: free of the Christian determinism which had decreed all bodily functions, lusts and sicknesses as signs implemented either by God or the Devil, emancipated towards the notion of identity and at the same time normalized and disciplined by an ever tighter growing network of societal rules, the body became one of the key features of cultural theory, while the natural sciences began to investigate its physical workings with a vengeance – up to the point when the borders between the 'wet ware' of the human body and the technology used to heal, mediate and even create it today begin to become ontologically obsolete.[100]

To talk of 'embodying' nowadays thus carries an ambiguous message, especially in a context where the question is asked whether or not to sent humans into space: on one hand, 'embodying' in the NASA motto may well be read as a tentative move towards using the human body in space exploration, on the other, the distinctly indistinct subject of the NASA missions – there is no 'I' or even 'we', however faintly, to be discerned in the motto, leaving the empty space to be phantasmatically filled with the amorphous identity of an institution with all its consolidated political and financial interests – doing the embodying might just as well cancel this possibility, as there seems no such thing like a human body to be sent.

Said ambivalence is an intrinsic part of the choice of words manifest in the NASA motto, and it shows up the danger inherent to the idea of embodying, as Lem diagnosed it in *Solaris*: whether human body or physical body in the Aristotelian sense, giving something a body in any case means defining a shape for it which in turn fixates its content. A typical example for this is the well-known phenomenon of the man in the moon: the need to fix a shape seems to be an overwhelming one, decorative in its effects, but also prone to shuttering our perception towards other possibilities, including that of shapelessness, of unbodiness – which might well be what we encounter in space if we should indeed ever meet other forms of life.

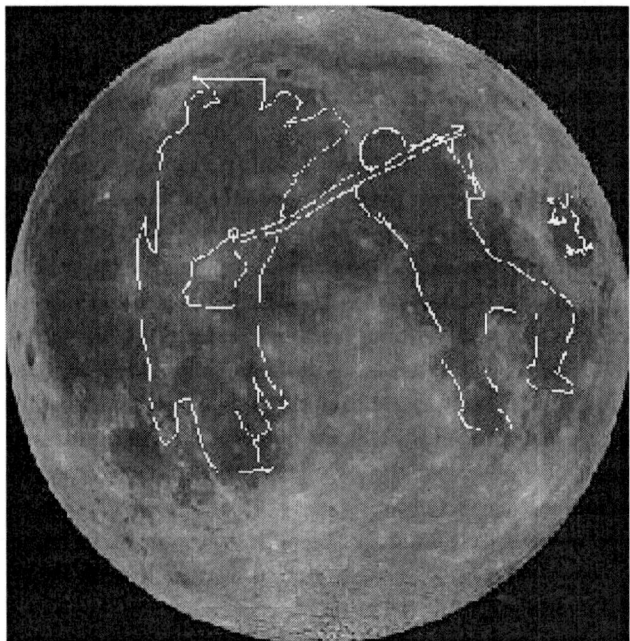

Fig. 3. *The Man in the Moon (source: www.planetfusion.co.uk).*

3.2.2.5. Spirit

At the beginning of the Genesis narrative it is God's Spirit that is hovering over the waters prior to, by God's word, becoming embodied in all the floral and faunal creatures due to inhabit the newly minted world. And, of course, it is the Spirit which later, in the Gospels, plays a major role in the mediation between God and humankind, even revising the Babel verdict by countering the multi-lingualism of God's people by giving the Apostles the ability to speak each and every language in the world during the Pentecost miracle. Again, the meaning of the word 'spirit' leans towards a progredient secularisation during the 18th century, when its religious implications are superseded by the inherently moral intellectuality of Enlightened man – but still, more than the other concepts, the term 'spirit' retained its biblical sense, at least in part, investing the idea of discovery here with the charisma of a God-given will to know for knowledge's sake.

This charisma, however, with the beginning of human space exploration has been invested with a particular kind of doubt, i.e., that of whether spirituality as established by religion and intellectually and morally reconfigured by Enlightenment will bear the strain of crossing the borders into space. The answer, which Roy Bradbury gives to this question in his 1951 short story collection *The Illustrated Man* is an emphatic 'no'. The following is an extract from the second of these stories, entitled *Kaleidoscope*. A spaceship has exploded on collision with a meteor, and its crew has been "thrown into space like a dozen silver fish" in their space suits. As they float apart from each other, radio transmission between them becomes weaker and weaker, until there is nothing but silence: "The many good-byes. The short farewells. And now the great loose brain was disintegrating. The components of the brain which had worked so beautifully and efficiently in the skull case of the rocket ship firing through space were dying one by one; the meaning of their life together was falling apart. And as a body dies when the brain ceases functioning, the spirit of the ship [...] was dying. [...] The voices faded and now all of space was silent. [...] They were all alone. Their voices had died like echoes of the words of God spoken and vibrating in the starred deep. [...] the shards of the kaleidoscope that had formed a thinking pattern for so long, [was] hurled apart."[101]

Of course, this is a literary text, and its use of metaphors to create an artefact is so obvious that there is no reason at all to consider the answer which Bradbury gives binding – but just along the same lines the evocation of the spirit in the NASA motto in its content is fragile at best, maintaining an integrity of the concept 'spirit' which has been developed under historical conditions that did not take the crossing of humans into space into account. This is why this passage metaphorically connects the 'spirit' with the metaphor of the kaleidoscope: encountering the

Other in space, even if it this means 'only' encountering virtual spatial endlessness, may, perhaps even must, mean the shattering of thinking patterns like that of the name of 'spirit'.

3.2.3. Conclusions

What the NASA motto gives to its readers is what Richard Geertz would have called a "thick description"[102] of space exploration – a description which is deeply imbedded in cultural contexts which of themselves are not explicitly named but implicitly drawn upon to make space exploration 'visible'. What becomes visible too is that 'the universe' as depicted by this 'thick description' is biased with atopological rather than ontological angle: the five concepts mentioned in the motto together constitute nothing less than a comprehensive topology of the discourse on humanity during the last centuries, extrapolating from these topoi a universe which is a priori anthropomorphous. Talking and writing about space using the semantic means offered by language in all the contextual and historical ramifications of connoted meanings must thus be seen as a process of constructing our object – as something which shows us first and foremost ourselves as we are, and so also shows us that we cannot see anything else but what we can see, what we are equipped to see, both technologically and epistemologically.

My analysis neither laments nor denounces the essentially narcissistic quality of our perception of space or the not less significant colonizing attitude towards the Other which springs from it. It is a mere banality to state that a medium as developed by humans, as language undoubtedly is, is at all levels structured to mirror and feed back into the identity of the species whose physical and mental abilities provide the condition of its existence. It might, however, be interesting to know whether whoever authored the NASA motto was fully aware of what he or she was doing in drawing on the subtexts and semantic reservoirs of meaning adherent to the terminology he or she employed, as the ad hoc of the medium, as opposed to its historical emergence, is indisputably a matter of choice: if the author knew what he or she was doing, this would mean that the employment of those five concepts was strategically used to ensure an uncritical identification with the agenda of NASA missions into space, considering that said agenda are coached in the immediately recognizable conceptual pillars of human identity on Earth. If, on the other hand, the author drew on this recognizability because he or she simply judged the terms employed as conveniently covering the issues concerned in a less-than-complex way suitable to the intellectual abilities of the average consumer, this would argue for an uncritically affirmative anthropomorphous projectionism. The latter might induce some doubts as to the institution's critical potential necessary for a responsible stance towards the issue concerned, while the former argues for an in-depth knowledge of human susceptibility for well-staged manipulation tactics and thus falls under the heading of public relation functionaries' strategic cynicism, inviting and cementing trust in NASA's professionalism.

The point of this sketchy excursion into what has long since been renounced by cultural studies as the 'author's intention' is not that of ideological polemics but rather that of emphasizing that the conscious knowledge of the fact that we do not really know precisely what we talk or write about when we mean 'humans in space' can be used productively in dealings with phenomena 'out there', from the mental and emotional impact of long-term existence of humans in, say, colonies on Mars to the possible contact with non-human life. In the case of the NASA motto, if we adhere to the notion that its author knew precisely what he or she was about, this productive use is documented in the supremely well-staged invitation to bring our own notion of humanity to bear on our perception of the Other and thus stabilize our will to expand into space along familiar lines, rather than becoming prey to fears of what incalculable forces might sabotage our progress. These fears, as we know so little, being out of necessity no less irrational than the assumption of an altogether anthropomorphous universe.

This leads me to my final resume of my thesis: missing the fact that it is impossible to construct anything like an accurate image of 'the universe' means, first and foremost, missing the chance inherent in the knowledge of this very impossibility. The Humanities have been long since aware of the problems connected with the limits, patterns and self-reproducing conceptionalizations of our perception. They do not offer any standard recipe for dealing with those problems. On the contrary, they make us aware that it is precisely the notion of standard recipes which make us vulnerable. What may be and indeed must be gained from employing the humanities in the development of the world-wide project 'Humans in Space' is the development of a double-edged awareness of the problem: first, the awareness of that how we talk and write about space always has an artificial, even fictionalizing dimension which we cannot avoid – but that we can certainly avoid not taking this into account; second, the awareness that what we may meet out there cannot be a priori presumed to be anything like ourselves, neither in knowledge or learning, in body or spirit.

The consequence of this line of thought is that when we design a one-page summary on the reasons for sending humans into space for the U.S. president, we need to make a case not only for the probability of surmounting the financial and technological limits which yet keep humanity from existing in space, but also for the awareness that the distance between humanity and the Unknown Out There can only be breached by what Homi Bhabha called "the borderline work of culture":[103] whatever happens when humans leave Earth to live and work in outer space will happen not within what our historically grown notion of humanity has come to treat as a given framework, but between this framework and potentially wholly different others, including both the challenges of a new environment with the ensuing necessity of redefining humanity itself and the (however yet improbable) contact with other forms of life. The awareness of this may in time well prove to be the crucial asset of humanity and, by methodological inference, of Humanities in Outer Space.

[89] Huang, Michael. "The Top Three Reasons for Humans in Space". The Space Review 11 Apr. 2005 http://www.thespacereview.com/article/352/1.

[90] "Since hazards from asteroids and comets must apply to inhabited planets all over the Galaxy, if there are such, intelligent beings everywhere will have to unify their home worlds politically, leave their planets, and move small nearby worlds around. Their eventual choice, as ours, is spaceflight or extinction." Sagan, Carl. Pale Blue Dot: A Vision of the Human Future in Space. New York: Random House, 1994. p. 327.

[91] McLuhan, Marshal. Understanding Media: The Extensions of Man. Cambridge: MIT Press, 1994. p. 7.

[92] "Current Missions". Website of NASA. http://www.nasa.gov/lb/missions/highlights/index.html

[93] Knorr-Cetina, Karin D. Epistemic Cultures: How the Sciences Make Knowledge. Cambridge, MA: Harvard University Press, 1999.

[94] Dünne, Jörg, ed. Raumtheorie. Grundlagentexte aus Philosophie und Kulturwissenschaften. Frankfurt am Main: Suhrkamp, 2006. p. 26; as for the modification of the passage towards a mere simile in the context of ongoing scientific debates.
Koyré, Alexandre and Bernhard Cohen. "The Case of the Missing Tanquam: Leibniz, Newton and Clarke". Isis 52 (1961): 555–566.

[95] Gen. 11, 4. – For the following quotations from the text of the Bible cf. the King James Bible text version at http://www.bartleby.com.

[96] Gen. 11, 6.

[97] Gen. 11, 7.

[98] The full title of Athanasius Kircher's book is as follows: Turris Babel, Sive Archontologia Qua Primo Priscorum post diluvium hominum vita, mores rerumque gestarum magnitudo, Secundo Turris fabrica civitatumque exstructio, confusio linguarum, & inde gentium transmigrationis, cum principalium inde enatorum idiomatum historia, multiplici eruditione describuntur & explicantur. Amsterdam: Jansson-Waesberge, 1679.

[99] Immanuel Kant's short essay 'Was ist Aufklärung?' appeared in December 1784 in the Berlinische Monatsschrift (Berlin Monthly); for an English translation cf. http://www.english.upenn.edu/~mgamer/Etexts/kant.html.

[100] Cregan, Kate. The Sociology of the Body: Mapping the Abstraction of Embodiment. London: Sage, 2006. Burkitt, Ian. Bodies of Thought Embodiment, Identity and Modernity London: Sage, 1999. Csordas, Thomas J, ed. Embodiment and Experience: The Existential Ground of Culture and Self. Cambridge: Cambridge University Press, 1994.

[101] Bradbury, Roy. The Illustrated Man. London: Grafton, 1977. p. 43.

[102] Geertz, Richard. "Thick Description: Towards an Interpretive Theory of Culture". The Interpretation of Cultures. New York: Basic Books, 1973. pp. 3–30.

[103] Bhabha, Homi K. The location of culture. London: Routledge, 2000, p. 16.

3.3 Towards a new inspiring era of collaborative space exploration

Nicolas Peter

3.3.1. Introduction

In recent years, space exploration has topped the agenda of most space-faring countries. This interest in solar system exploration can be illustrated by the development of the U.S. Vision for Space Exploration, the European Space Agency (ESA) Aurora programme, as well as robotic exploration missions under development in India, China, Japan and Russia. This is completed by the fact that there is now a new context in space affairs resulting from a new international landscape after the end of the Cold War that has opened an opportunity for the definition of a new framework for international relations in space. Furthermore, recent geopolitical developments, combined with the funding constraints of the various space-faring countries, have made it clear that greater international cooperation will be important for major future space activities. This is particularly true for a long-term space exploration programme due to the increasing complexity of such missions. However, just as important as the destination, the journey, when done as an international cooperative effort, will be able to inspire and motivate a wide international community to back mankind's next grand challenge and open a new phase of space exploration.

3.3.2. How are current space exploration plans different from earlier space endeavours?

The history of space exploration can be structured in four phases, each having distinct features and characteristics as illustrated in Figure 4.

The first phase of space exploration corresponds to the "Proto-space Age" during which major advancements in the field of rocketry and astronautics were made before the Second World War under the leadership of visionary individuals such as the American Goddard, the German Oberth and the Soviet Tsiolkovsky (Figure 4).

The second phase of space exploration or "Space Exploration 1.0" took place during the Cold War from the late 1950s to the late 1980s (Figure 4). For more

Exploration phase	Time period	Characteristics
Proto-space Age	before WWII	Leadership of individuals (and societies) such as Robert H. Goddard, Konstantin Tsiolkovsky, Hermann Oberth, Hermann Potocnik, Robert Esnault-Pelterie, etc. influenced by Herbert G. Wells, Jules Verne and other science fiction authors
Space Exploration 1.0	Cold War	Competition between the United States and the U.S.S.R. with cooperation limited to intra-blocs partnerships driven primarily by political reasons (duopoly situation)
Space Exploration 2.0	1990s–now	Exploration plans with new actors willing to participate driven primarily by scientific motives (oligopoly situation)
Space Exploration 3.0	soon	Era of participatory human exploration (States, industries, universities and others non-governmental organizations) driven primarily by the quest for knowledge (including Arts and Humanities disciplines) as well as economic potential (open-market situation)

Fig. 4. *Classification of space exploration era (source: Nicolas Peter).*

than three decades, space was viewed as one of the areas for peaceful Cold War competition between the United States and the U.S.S.R. as a substitute for armed conflict. Space exploration in this period was an emblematic element in this rivalry, as illustrated by the number of exploration missions (Figure 5). But, in the context of the "space race", international cooperation was central to the two space powers' political strategy for similar reasons (demonstrating leadership ability and technological capabilities). Therefore, "intra-bloc" cooperation was the norm.[104] The

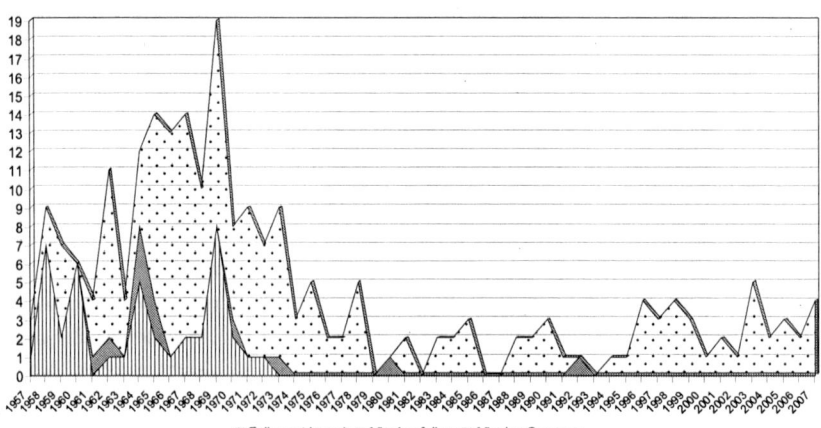

Fig. 5. *Number of space exploration missions over time (source: Nicolas Peter).*

first phase of space exploration was driven by Cold War rivalry with cooperation extended to political allies of the two principal space powers. During this era space exploration was thus limited to a small number of international science missions consisting of in-kind contribution and ad-hoc coordination mechanisms (i.e., 1975 Apollo–Soyuz Test Project).[105]

The third phase of space exploration, or "Space exploration 2.0" started in the 1990s as a result of the changing space context in the post-Cold War era (Figure 4). The Cold War and its East versus West political environment had evolved from a bipolar space world dominated by the United States and the U.S.S.R. into a multipolar world characterized by the rise of many new actors with increasing technical capabilities, such as Europe through the ESA and other national space agencies like that of Japan.[106] This period can, moreover, be characterized by an internationalization of space activities in which the number of space agencies in the world has been steadily rising since the 1990s and reached 36 in 2005 (Figure 6).[107] This multiplication of space agencies in the post-Cold War context is also completed by an emerging globalization of the space actors, with space agencies now scattered all over the world, as they are not only limited to the "North" with new institutions being created on all continents.[108]

This growing number of countries with varying ranges of space capabilities is leading to an increase in the options for cooperation. As a result of this changing geopolitical space context, bilateral and multilateral agreements between agencies have been growing, particularly in space sciences (Figure 7). This is obviously resulting in a growing pool of potential partners for international space exploration cooperation initiatives.

In this context, at the difference of the second phase of space exploration (Space Exploration 1.0), the third phase of space exploration (Space Exploration 2.0) has seen an increasing number of space agencies being involved in space exploration missions (Figure 8).

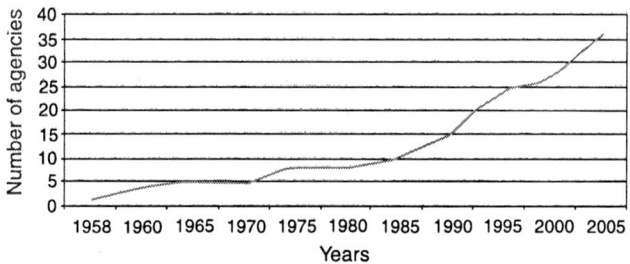

Fig. 6. *Evolution of the number of national civilian space agencies over time (does not include multinational space agencies)[108] (source: Nicolas Peter).*

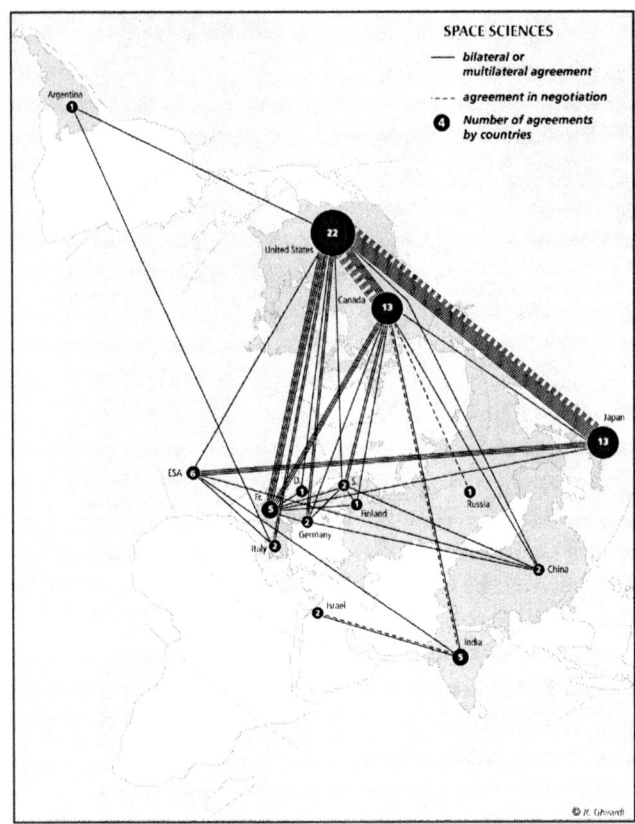

Fig. 7. *Cooperation in space sciences among major space agencies between 1992 and 2004[109] (source: Nicolas Peter).*

Space exploration phase	Countries leading space exploration missions	Number of missions
1957–1989 Space Exploration 1.0	U.S.S.R.	112
	United States	75
	Japan	2
	ESA	1
	Total	**190**
1990–2007 Space Exploration 2.0	United States	26
	ESA	7
	Japan	5
	France	1
	Russia	1
	China	1
	Total	**41**

Fig. 8. *Space agencies involved in the two phases of space exploration (source: Nicolas Peter).*

The changing geopolitics of space activities is completed by the increasing experience of long-term cooperation in space exploration activities, and in particular the International Space Station (ISS) experience where the research facility currently being assembled in space is a joint project of the United States (NASA), Russia (Roskosmos), Japan (JAXA), Canada (CSA) and Europe (ESA).

As already mentioned, all of the major space-faring countries have launched space exploration missions in recent years and have shown different degrees of interest in solar system exploration. Many robotic missions are already planned or

Country	Mission	Approximate launch date	Lander/orbiter
United States	LRO (RLEP1)	2008	Orbiter
United States	LCROSS	2008	Lander (impactor)
India	Chandrayaan-1	2008	Orbiter
Russia	*Mission to Moon*	2009	Orbiter
China	Chang'e-2	2009	Orbiter
India	Chandrayaan-2	2011	Lander + Rover
United States	LADEE	2011	Orbiter
United States	GRAIL	2011	Orbiter
Russia	*Moon Lander*	2011	Lander
Russia	LunaGlob	2012	Orbiter + surface probe
United Kingdom	MoonLITE	2012	Orbiter + penetrators
Germany	German national mission	2013	Orbiter
China	*Chang'e-3*	2013	Lander + Rover
Japan	Selene-2	2013	Lander + Rover + Impactor
United States	Lunar Lander 1&2	2014	Lander
United States	Robotic Lunar Mission	2015	*Not defined yet*
Japan	Selene-3	2017	*Sample Return*
China	Recoverable Moon Rover	2017	Rover
United States	*Lunar Lander 3&4*	*2017*	*Lander*
India	Chandrayaan-3	2019	*Sample Return*
United States	Manned landings (several)	2020+	Human lander
South Korea	*Lunar Orbiter*	*2020*	*Orbiter*
South Korea	*Lunar Lander*	*2025*	*Lander*

Fig. 9. *Planned lunar missions (in italic tentative missions) (Source: Nicolas Peter).*

underway. For instance, following the footsteps of ESA's SMART-1 (Small Missions for Advanced Research in Technology-1) orbiter, a fleet of automated spacecraft are currently being dispatched to the moon by China and Japan and will soon be joined by the United States and India (Figure 9). At the same time, several robotic spacecraft will also be sent to Mars and other planets in the solar system.

The recent catalyst for this movement is President George W. Bush's bold redirection of the U.S. civilian space programme to pursue the exploration of the Moon, Mars and the "worlds beyond".[111] Consequently, since the proclamation of President Bush's Vision for Space Exploration in 2004, NASA's activities have been driven by the goal to return humans to the Moon before 2020. However, the new U.S. space exploration policy also seeks to "promote international and commercial participation in space exploration to further U.S. scientific, security, and economic interests" and invites "other nations to share the challenge and opportunities of this new era of discovery".[112] In this context, following the reorientation of the major civilian space agency in the World, NASA, in the four years since the announcement of the U.S. Vision for Space Exploration, many countries have expressed an interest in collaborative exploration programmes. Informal discussions on goals, capabilities, and timelines for future space exploration, particularly focusing on the moon have taken place among major space agencies illustrating the paradigm shift in space exploration whereby international cooperation is becoming central to any long-term space exploration strategy.

In particular, as a result of the work between representatives of 14 space agencies, which have met four times since August 2006, on 31 May 2007, at the third ESA/ASI workshop on "International Cooperation for Sustainable Space Exploration", a 25-page report "Global Exploration Strategy – The Framework for Cooperation" was released as the first product of an international coordination process among those agencies.[113] From this document, the international definition of space exploration may be read as "a global, societal project driven by the goal to extend human presence in Earth–Moon–Mars space" with the five explorations goals being: human missions to near Earth orbits, robotic and human exploration of the moon, human missions to liberation points of the Earth–moon and Earth–sun systems; robotic (and human) exploration of near-Earth objects (NEOs); robotic and human exploration of Mars. The document also discusses the rationale for society to explore space based around five major themes: new knowledge in science and technology, sustained presence – extending human frontiers, economic expansion, a global partnership, and inspiration and education. Therefore, the May 2007 document illustrates the awareness of the value of space exploration as a global, societal project. The joint document is also supported by a large database of possible exploration objectives spanning the whole spectrum from hard sciences to economics and social benefits.

The "Global Exploration Strategy" develops also the case for globally coordinated space exploration and investigates, among others things, a framework for the future coordination of global space exploration. It recognizes that "Sustainable space exploration is a challenge that no one nation can do on its own. We are now entering a new wave of space exploration, one of historic significance. The United States has developed its Vision for Exploration, the European Space Agency has its Aurora space exploration programme. China, India, Japan and Russia have ambitious national projects to explore the Moon or Mars, while future national missions are being discussed in Canada, Germany, Italy, Republic of Korea and the United Kingdom".[114] This strategy is designed to introduce minimum standards of interoperability to facilitate cooperation, while permitting individual countries to pursue their own national strategies. Following the adoption of those aforementioned basic principles, the 14 signatory agencies set as the next step the creation of a "Coordination Mechanism" in the form of a semi-permanent body aiming to coordinate further steps in harmonizing the exploration effort. The formal establishment of an "International Space Exploration Coordination Group" for steering the further implementation of the international coordination process, with the terms of reference has been agreed at the level of the directors of participating space agencies. This group aims therefore to facilitate the exchange of information on space exploration plans.

While new space agencies plans are currently underway, we are however at the verge of transitioning to a new space exploration era or "Space Exploration 3.0" that will be an era of participatory human exploration (Figure 4). This new phase of space exploration resulting form an organic evolution will involve unlike the previous two space exploration era (Space Exploration 1.0 and 2.0) not only states through their space agencies, but also industries, universities and others non-governmental organizations. While other initiatives were primarily driven by foreign policies motives and technology development purposes, this new adventure will be driven primarily by a quest for knowledge, involving not only the hard sciences but also the Humanities and Social Sciences. Economic potential for space exploration will also increasingly become a driver for long-term plans, because until now space exploration has driven and funded largely by governmental actors. This is already beginning to change as entrepreneurs start to play a significant role in the utilization of space especially through a series of U.S. led entrepreneurs' initiatives such as the Google Lunar X Prize presented in September 2007. The X Prize Foundation and Google Inc. announced a new cash prize competition aiming to start a commercial race to the moon with 30 million U.S. dollars in incentives. The goal of the new prize is to land a privately funded robotic rover on the moon that is capable of completing several mission objectives, such as roaming the lunar surface to a distance of at least 500 m and transferring a set of specific video and images back to Earth. This initiative aims thus to open a new era of lunar exploration and to extend the economic sphere to the moon by harnessing the existing untapped innovative power around the world. Following such an initiative, it is thus likely that private enterprise will play an ever-increasing role in future exploration programmes. This future expansion of entrepreneurial activities into space is of particular importance, because it will offer a broad range of new opportunities and contribute to enhancing the sustainability of space exploration plans and programmes.

3.3.3. Benefits of international cooperation

As aforementioned, since the pioneering of space activities in the late 1950s, international cooperation[115] has been a central element of the strategy of most countries involved in space activities including space exploration. International cooperation had been used to expand not only their technical and scientific capabilities of countries, but also their political ties. For instance, during the second phase of space exploration (Space Exploration 1.0), cooperation was used to expand political influence over allies from the two superpowers respective blocs. In the current geopolitical context, the arguments in favour of cooperation have not changed fundamentally since the dawn of the "Space Age" they are still a combination of scientific, economic, political and security motives.

The benefits of international cooperation are numerous and well documented. Among others, they include improving capability, sharing costs and building common interests and increasing the total level of available resources, eliminating the duplication of efforts, and improving international relationships.[116] International cooperation in space activities allows rationalizing and optimizing resources and mounting missions that would otherwise not be possible. It is therefore generally conceded that international cooperation expands the scope of programmes beyond the individual participants' capabilities by tapping into the resources of multiple countries and enlarges the spectrum of possible missions. This expansion of resources made available through cooperation is not only just financial, but also scientific and technological. International cooperation enhances also domestic legitimacy to space projects and gives them internationally credibility and makes them also less vulnerable to cancellation due to domestic political or financial problems.[117] It is now an integral part of the space policy and strategy of the different space agencies around the world and countries do no longer initiate or carry out a significant space programmes without some element of foreign participation. International cooperation can therefore be seen as a critical enabler and one of the building blocks for any long-term space activities, and, in particular, exploration activities.

3.3.4. Metaprinciples for space exploration

While international cooperation is seen as a critical enabler for space exploration plans, it also carries risks. For instance, it is recognized that international cooperation adds layers of complexity to the design and management of programmes, and also affects successful budget and schedule performance. Furthermore, states generally cooperate when it benefits their self-interests and therefore, partners may be pursuing common programmatic goals, but for different reasons, as each partner's space programme exists within its own political environment.[118] In this context, since not all countries regard international cooperation equally and pursue collaborative endeavours for the same motives, enduring space exploration architectures require, as underlined by Correll and Peter, that *metaprinciples* for international exploration programmes be followed. In particular, to be successful a long-term space exploration programme will need to encompass the following series of major *metaprinciples*:

• Any long-term space exploration programme will need to rely on an open-systems approach providing flexibility to leverage cooperative opportunities. The most well-known example of open-systems architecture is the Internet.
• Space exploration programmes will need to be robust to sustain any potential failure that may arise.
• The exploration plan will need to be affordable and adequately financially planned.
• Any long-term space exploration endeavour will have also to be visible to be sustainable. The exploration strategy therefore needs to be multifaceted and inspirational to involve a broad stakeholder community. Long-term exploration programmes will thus need to consist of a mix of robotic and human missions to be successful. This will allow covering a wide range of interests from the excitement of human spaceflight to the quick pace of robotic spacecraft for technological and scientific purposes.
• Following the likelihood of the changing geopolitical future, any long-term exploration programme will need to rely on international partnerships as no one has the means to do it alone. Therefore, international cooperation must become an anchor of any long-term strategy.

3.3.5. Inspirational potential of international cooperation

Space exploration encompasses a complex set of activities and offers opportunities for broad international engagement and participation. Sustained human missions beyond low Earth orbit will not be possible with the resources of a single country. Moreover, because of the multiplication of space-faring countries and newcomers in space with increasing capabilities, there is a growing number of possibilities for cooperation. Space agencies around the world are now looking to a variety of partners as they plan their future endeavours. There will thus inevitably be opportunities for many other countries to make major contributions to a global programme. Future space exploration endeavours will therefore involve significant collaboration between space-faring countries, but also with newcomers.

In today's time of globalization and rapidly shifting international relationships it is essential that various countries unite towards a grand and peaceful goal. International cooperation in space exploration represent one of the most efficient and visible ways of affirming a willingness to cooperate with others. Space exploration, and human spaceflight, have been used from the onset by the superpowers as a means of impressing the world. Thus such high profile activities represent one of the most efficient and visible ways of affirming an assertive global position in a peaceful manner and demonstrating a willingness to cooperate with other countries.

Curiosity about what other worlds are like inspires public interest in space exploration. Questions about the nature and the origins of life have fascinated all cultures throughout history, and mankind is now embarking to an Odyssey that ultimately will be able to fulfil these elements. However, while space exploration and human spaceflight are the most emblematic aspects of the space endeavour, they remain far removed from the public's everyday consciousness. It is therefore necessary to raise the general public level of interest in space exploration and to foster the exploration culture across generations and nurture public constituencies for long-term space exploration. Any of such endeavours will motivate the public, but if done in international cooperation, it can also unite humanity. International cooperation will therefore be crucial in this regard. Space exploration (particularly human space exploration) is a source of inspiration permitting to foster excitement and encourage discovery in a cooperative and international fashion. It will allow assembling humanity behind a peaceful goal and will facilitate increasing the level of international involvement in space exploration ventures.

Space exploration could thus inspire countries to work together for a common purpose. However, future endeavours will not only be restricted to current space-faring countries because they will transcend all disciplines and not only consist of scientific and technological-related activities. It will therefore offer new possibilities of involvement even for currently non-space-faring countries. Every country is likely to be involved in any long-term human space exploration efforts, since this will attempt to establish an enduring human presence in the solar system. Therefore, as the U.S., Europe and other countries embark on new explorations plans and programmes, they should lay the foundations and establish precedents that invite a host of participants and followers, because one of the major benefits of those exploration endeavours more than the destination and related discoveries will be the journey itself, as international cooperation will allow fostering broad public support across countries. Thus, irrespective of the destination, each journey has the potential to capture the imagination of the general public and inspire future generations of scientists and engineers.

International partnerships will enable countries to develop a common understanding of their respective interests, to share lessons learned and demonstrate goodwill. International space exploration initiatives will thus allow increasing cultural awareness and improved amity and fraternalism, therefore providing direct Earthly benefits.

3.3.6. Conclusions

Since the announcement of the U.S. Vision for Space exploration in 2004, space exploration has become again, like, in the Cold War a major element of the strategy and plans of major space-faring countries. However, unlike the earlier period, the post-Cold War context is undergoing a rapid evolution with a growing number of new actors considering and engaging in space exploration activities. However, all the existing and emerging space powers have made the decision to engage in space exploration, mainly robotic missions, while human exploration is a central element of the exploration plans of major space powers due principally to its emblematic nature.

Throughout its existence, mankind has been driven by a desire and a drive to explore. Space exploration beyond Earth orbits will thus definitively be one of the challenges of the 21st century. It will allow humanity to assemble behind a peaceful goal since space exploration is mankind's next grand challenge. Humanity is therefore on the threshold of stepping off into space for a unique Odyssey that may lead to the discovery and exploration of new worlds, because as Konstantin Tsiolkovsky said "The Earth is the cradle of humanity, but mankind cannot stay in the cradle forever."[119] However, to ensure the sustainability of a long-term exploration plan is the necessity to inspire a broad constituency base. And, this should be done at two levels: "intra-countries" to foster broad public engagement and "inter-countries" as any long-term exploration programme is currently beyond the capabilities of any individual country. International cooperation will therefore be crucial. However, because major societal and political changes will undoubtedly take place in the course of any long-term space exploration, *metaprinciples* for space exploration should be followed. These include architectural openness and flexibility, visibility and affordability. Moreover, while sciences lie at the core of space exploration, public support is a vital *metaprinciple* to ensure the viability of long-term exploration plans and international cooperation is an area that particularly appeals to the general public.

It is also of paramount importance not to oppose a utilitarian and exploratory vision since the future of space will be a utilitarian exploration. It is thus expected that the future of space exploration will evolve into a fourth phase of space exploration, the so-called "Space Exploration 3.0", that will be international, human centric, trans-disciplinary and participatory, and will provide an opportunity to inspire, motivate and involve an ever increasing number of countries. Unlike the previous two space exploration periods, it will involve not only states and space agencies, but also industries, universities and other non-governmental organizations. This adventure will be driven primarily by a quest for knowledge, involving not only the hard sciences but also the Humanities and Social Sciences, as well as by economic potential, thus increasing the possibilities to include a variety of space actors, even emerging ones.

[104] Peter, Nicolas. "The Changing Geopolitics of Space Activities". Space Policy 22 (2006): 100–109.
[105] Ibid.
[106] Ibid.
[107] Ibid.
[108] Ibid.
[109] Ibid.
[110] Ibid.
[111] However, even before the announcement of this new U.S. initiative, space agencies around the world were developing plans for robotic and human exploration missions beyond Low Earth Orbits (LEOs).
[112] "The Vision for Space Exploration". NASA. Feb. 2004 www.nasa.gov/pdf/55583main_vision_space_exploration2.pdf.
[113] The 14 agency signatories are the national space agencies of Australia, Britain, China, Canada, France, Germany, India, Italy, Japan, Russia, South Korea, the United States, Ukraine and the 17-country ESA.
[114] "The Global Exploration Strategy Framework: Executive Summary". NASA. 31 May 2007. www.nasa.gov/pdf/178109main_ges_framework.pdf.
[115] For the purpose of this article "international cooperation" is used as a general term denoting international governmental participation in a project; industrial and commercial cooperation are not discussed.
[116] Correll, Randall R., and Nicolas Peter. "Odyssey: Principles for Enduring Space Exploration". Space Policy 21 (2005): 251–258.
[117] Ibid.
[118] Ibid.
[119] Tsiolkovsky, Konstantin. Brainy Quote. http://www.brainyquote.com/quotes/authors/k/konstantin_tsiolkovsky.html.

CHAPTER 4

FIRST ODYSSEY:
HUMANS IN EARTH ORBIT:
WHAT EFFECT DOES IT HAVE?

4.1 Summary

Marcel Egli

Let us leave for the first odyssey, breaking terrestrial boundaries and explore Earth's Orbit. This is not an extraordinary task anymore and a few hundred people (astronauts, cosmonauts, taikonauts, etc.) already had the privilege to go there. Currently, we are on the brink of having commercial transportations available for ordinary people who wish to experience the breathtaking view of Earth from an altitude of at least 100 km and to feel the effect of microgravity. With such travelling possibilities in view, there will be far more people in the near future who will be able to report on the sensation of being in Earth's Orbit. Due to the fact that trips to Low Earth Orbit (LEO) are carried out regularly by the major space agencies, this first odyssey gives a reflection on the experiences gained so far in order to pave the way for the next journeys heading for outer space, where no human being has ever been before.

As spectacular manned space missions are, the roots of space flights are less glamorous. The first modern rockets were designed and built by armed forces in order to gain advantages in combat. After the war, that technology was further developed, driven mainly by the race between the two superpower nations U.S. and Russia for dominance in space. A particular impulse for the U.S. space program was the "Sputnik shock". In response to that event, President John F. Kennedy announced in 1961 an ambitious space program to send humans to the Moon within a decade. Indeed, U.S. astronaut Neil Armstrong became the first man who walked on the Moon nine years later.

In the past, brave adventurers were driven to travel to the unknown in prospect of glory, honour and prosperity for their sponsors. Similarly, space travels were chiefly undertaken for demonstration of technical predominance and power. The Apollo program impressively underscored the leading status of the U.S. technologies at that time. Today, the technological advantage of the U.S. is less prevailing and nations are cooperating together on space programs. The driving forces of the past have disappeared. Nevertheless, new goals for manned spaceflights have been set. The declared long-term aim of several space agencies is to return to the Moon and even visiting Mars, after completion of the International Space Station (ISS). Space activities in Earth Orbit are the stepping stone for space exploration, starting with flights to the Moon and the erection of human settlements. All the experience gathered with LEO space activities like maintaining the permanently manned

Fig. 1. *Camille Flammarion, L'Atmosphere: Météorologie Populaire (Paris, 1888) (source: Wikimedia).*

space station ISS, are critical for the successful operation of future human spaceflights.

ESA astronaut Claude Nicollier introduces us to aspects of LEO activities "with the eyes of an astronaut". He was a member of the first group of ESA astronauts selected in 1978. Shortly after, he joined *Group9* of NASA astronauts for Space Shuttle training at Johnson Space Center in Houston, Texas. During four Space Shuttle missions (STS-46, STS-61, STS-75, and STS-103), he spent more than 1000 hours in space including a spacewalk of more than 8 hours. His explanations make clear that it is possible to perform technically demanding tasks in space, already now and there should be no hesitation in going further for what technology is concerned. However, there are clear limits for a fragile human body in space and he stresses the importance of robotics for a successful human exploitation of space. But probably the most impressive experience for an astronaut in Earth Orbit is the absence of national boundaries. Regardless of the nationality, culture, and political background, multicultural teams of astronauts are working together nowadays to successfully complete the mission. The quintessence of Claude Nicollier's contribution is that human space flight is the discipline of choice to bring people together. Let us hope that the manned space program serves as a paradigm to demonstrate how people from different cultures can live and work together.

Space flight programs are ultimately linked to the latest technological achievements and often trigger inventions which might find the way to our modern life (e.g. telecommunication). Richard Tremayne-Smith, an expert in space engineering from the British National Space Centre points out in his chapter that in manned space missions, technology defines the limitations but at the same time also represents the enabling capability for space exploration. Therefore, all efforts need to focus onbreaking limits which then automatically lead to new inventions. This, in turn, is a warrant of technological progress and economical growth on Earth.

However, technological development needs to take into consideration other factors. Gabriella Cortellessa from the National Research Council of Italy, Institute for Cognitive Science and Technology, advises in her contribution that care must be taken in designing potentially overwhelming technologies. A certain degree of freedom needs to be maintained in order to allow man and machine to cooperate. Her approach is that the interaction between intelligent technologies and humans in space should be a form of collaboration rather than a master–slave interaction or completely technology driven.

Not only will the technological aspect of our first odyssey bring us to the frontiers of our daily life but also to the frontier of legal affairs. The construction of an international space station was and still is an ambitious undertaking which created may novel situations. In the past, manned space activities were of relatively short duration and often conducted by one nation only. Therefore, only a few legal agreements had to be put in place. But with the building of a permanently manned station operated by international teams, new legal rules and contracts had to be negotiated. Frans G. von der Dunk from the International Institute of Air and Space Law, Leiden University, Faculty of Law explains in his chapter the historical background of the legal basis on which the space activity is based on and discusses the implementation of these rules on the example of ISS.

The insights of these four experts introduce us to various aspects of space activities in Earth Orbit. Most of it is based on the experiences accumulated during almost 50 years of the human spaceflight program. Nevertheless, these experiences are very important in shaping future space missions which will bring humans to outer space.

4.2   With the eyes of an astronaut

Claude Nicollier

4.2.1. The discovery

Humans have ventured into space for many different reasons since the first flight of Yuri Gagarin on April 12, 1961: Human spaceflight has changed from a way to demonstrate power and capabilities in the Cold War climate of the 1960s, it has now turned into a tool for science, technical development. It is a demonstration of our capability to adapt and be highly productive in an environment totally unlike the one life has evolved in for billions of years on the surface of the Earth. Not that a spirit of competition is totally absent from the much larger set of players than before, but solid partnerships have been created, without displacing factors such as prestige, national pride, and symbolism which, as we may expect, will always accompany such undertakings.

Despite the basically non-scientific nature of the Apollo Programme, what we have learned about the composition and history of our celestial neighbor from the six missions that reached the surface of the moon has been very valuable. What is more, it has certainly triggered the interest in using humans in space to expand the boundaries of knowledge in many areas, including space physics, astronomy, material science, biology, and physiology. All subsequent human space missions were restricted to low Earth orbit, culminating in the International Space Station program (ISS). These missions, using the Shuttle and Soyuz as transportation vectors, have definitely demonstrated that humans can live and work in space for months without any demonstrated negative health effects, provided they adhere to a strict physical exercise protocol while on orbit.

The Hubble Space Telescope (HST) servicing missions (four up to 2007) and the ISS Assembly missions have established the effective use of a combination of spacecraft-based robotics and extravehicular activity for the installation, removal or exchange of sometimes large replaceable units on a spacecraft, and sometimes for fine work requiring dexterous handling with the help of specially designed tools. This is not a small achievement, and is an important demonstration in view of future human exploration-type missions in the Solar System.

4.2.2. The view from above

On two occasions, I have had the privilege of being a crewmember on an HST servicing mission (SM-1 and SM-3A). The perfect combination of remote control from the ground (for normal scientific operations) and human interventions on-orbit (for repair, servicing and/or improvements) ensured the extremely successful completion of the programme. Shuttle-based servicing missions have allowed the correction, in two instances, of the loss of essential telescope capabilities (optical resolution in 1993, attitude control in 1999). Automatic and robotic functions alone would never have achieved what has been done thus far with HST, and this is an important lesson for the future. Not that it always will be possible to afford the luxury of human spacecraft-based servicing, but, whenever it is feasible, this approach will provide us with a lot of options, flexibility, and the capability for correction or even recovery from critical failures (Figure 2).

It is well known that views of the Earth and of the starry nights are very spectacular from a Low Earth Orbit. Although more a background than a subject of close attention during busy times on-orbit involving robotics and/or spacewalks, the Earth's surface and atmospheric phenomena receive a lot of attention and

Fig. 2. *Exchange of three Rate Sensor Units (RSUs) on HST, or HST SM-3A, December 1999 (source: NASA Picture).*

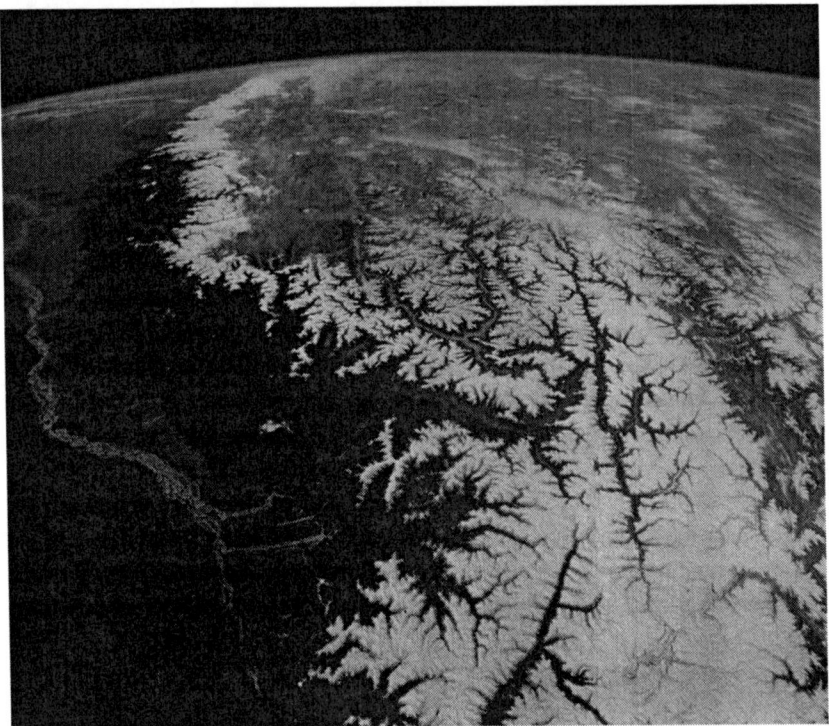

Fig. 3. *Himalaya mountain range from 300 km altitude, STS-75 (source: NASA Picture).*

recordings on digital cameras during quiet times on the Shuttle and the ISS. Most impressive are the large-scale geological features like the Sahara desert, the Himalaya mountain range and the Andes (Figure 3). The rapid succession of day and night, the beautiful lighting at every transition, the amazing spread of thunderstorm areas in the equatorial regions, the auroras, zodiacal light and the Milky Way, are unforgettable and are a substantial addition to the dimension of each expedition to Low Earth Orbit.

4.2.3. Where do we go next?

In the near future, Human Spaceflight will continue to develop along the lines of current programs – completion of ISS assembly and its exploitation (Figure 4), another HST servicing mission. A clear change will come about around the end of the next decade when the U.S. returns to the Moon, hopefully with some

Fig. 4. *Orion capsule approaching the Space Station (source: Lockheed Martin Corportation).*

cooperation with all or at least a fraction of the ISS partners. We can also count on advances in the field of human spaceflight by nations like China and India later. Hopefully, close cooperation between all of these nations will eventually become established.

Human spaceflight is a discipline of choice for bringing people together in an effort to expand the boundaries of knowledge. It is my sincere wish that Europe will also invest and make the proper choices to become a firm partner in the human component of future space exploration programmes.

4.3 Human spaceflight, technology development and innovation

Richard Tremayne-Smith

4.3.1. The first effect – inspiration from space

From ancient times, space has been a source of inspiration to man and he has looked for ways in which to relate man's existence to the ever changing canopy of the night sky. Over time, space, which exists beyond the fragile atmosphere that marks the outer reaches of normal human existence, has become increasingly important for human beings. Placing satellites into orbit around the Earth has allowed us to better understand our planet by observing weather and surface features as well as providing accurate timing and position information to support commercial and navigation services. Still, whatever else we derive from space in the way of services, it will always remain a continuing source of wonder and inspiration.

4.3.2. The second effect – supporting life on Earth

While technology helps us to take steps towards increasing our understanding and utilising our natural resources to improve the quality of life, we increasingly need to apply it to address environmental issues including climate change. The basic limits to growth that will result from the growing population and dwindling resources are best viewed from space where Earth, without boundaries, is the way astronauts see our blue planet. Space is firstly a tool to enable us to better understand the many processes that are taking place on Earth, but conquering space and being able to travel there in a sustainable and affordable way will be the key to supplementing terrestrial resources. To this end, the revived interest in space exploration and the genuine enthusiasm at the global level by space agencies is a welcome step in the right direction. The planning, preparation and general excitement can be shared by everyone just like the long-term benefits of human presence in space for the majority of people who will remain on Earth.

Fig. 5. *Comet Holmes in the night sky (source: Ian Morison).*[120]

4.3.3. Reality

It is not generally recognized that while the U.S. and Russia had developed a capability for travel beyond the Earth's orbit, and the U.S. even went on to land on the Moon, this capability was lost after the Apollo era. At present, we have no proven ability to send a person to the Moon. Therefore, it is important that when planning a return to the Moon and beyond, one should seriously consider and focus on how to build a capability that is sustainable over the long-term. Our presence in low Earth orbit should not be an end in itself, but a stepping stone on the way to broader human exploration of space.

There is a need to improve safety, especially in the area of human spaceflight, while at the same time, lowering overall transportation and infrastructure costs.

4.3.4. Technology and innovation

Learning from our experience of life in orbit and the initial sorties beyond, we should aim for some level of commercialisation as well as to increase knowledge that will be gained from greater access to outer space. Technology is our limitation as well as our enabling capability when venturing towards the stars.

129

Access to space is a very important area of technology development as we move towards the need for regular and safer travel into space, while at the same time, it is also of utmost importance to significantly reduce the costs. Space tourism is starting to touch on the price issue, as the only way to target a much larger audience is to bring down costs. At the bottom end of the market, this means a space experience, but not yet real access to space. Russia, for example, can still charge USD 20 million for a Soyuz flight and a few days at the International Space Station. The outlook of many more humans entering near-Earth space within a few years is high, and it is the human element that is driving innovation in this growing market.

Most importantly, as more countries join the new initiative for space exploration – making it a truly international endeavour – we will need to spend time to plan in advance how to manage the enlarged space environment, namely, *beyond Earth orbit*. We have already introduced significant volumes of rubbish, referred to as orbital debris, into Earth's orbit, which makes the use of these orbits more difficult and more costly than would have been the case if we had been better at preserving the near-Earth environment. We need to ensure that we have learned our lessons from this experience and keep other special regions of space as free from debris as possible. These orbits will include transfer orbits to the moon from Earth, the moon and its orbital environment[121] as well as the stable and other Lagrangian points of the Earth's moon and Earth's solar system. This area is the subject of ongoing work as part of the planning phase for the Global Exploration Initiative.[122]

Generally, there are international groups working on planetary protection based on the work by the Committee on Space Research (COSPAR).[123] Planetary protection is concerned with both forward and backwards contamination, attempting not to pollute other areas of space with earth-based contaminants and trying to ensure that returning space missions do not bring contamination back to Earth. *Over 1000 comets have been discovered by SOHO, the ESA/NASA Solar and Heliospheric Observatory. We are looking out for near-Earth-objects (NEOs), but what about the other areas of the solar system we may wish to travel to?*

4.3.5. What space has to offer

4.3.5.1. To humankind

Apart from the applications already mentioned, including support for the monitoring and maintenance of a sustainable Earth environment, we can look to a future

Fig. 6. *Comets approaching and disappearing into the Sun (source: http://science.nasa.gov/newhome/pad/ images/soho_corona.gif).*[124]

where space could provide the ability to remove the current limits to growth.[125] This does not mean that growth should necessarily be encouraged, although improvements in the quality of life or growth to reach new levels of future sustainability would be a positive development. Another way of looking at the issue is to consider the need to either eliminate the limits or stop growth. Space could contribute to helping to remove the limits, but only if long-term improvements to the Earth's environment are achieved.

We must also consider that most cultures have an element of growth built into their way of managing things, and who has ever heard of a best-selling business book entitled "How I managed a no-growth business". When we think about it, growth is relatively easy to manage, because the errors of hiring too many staff or leasing a factory that is too big can be covered up for a short time and presented as a good decision. Sustainability should be linked to a culture of real management where overall impact and not just short-term profits are considered. Later works by the original authors of the "Limits to Growth" has lead to an update to the original work called "Beyond the Limits to Growth". While making the point that many of the issues are now more critical, sustainability, it concludes, is still an achievable goal. When shifting the focus to the new challenges for humans in outer space, we must make sure to preserve the priority of sustainability of space operations as well as Earth-based sustainability.

Future prospects for space to support long-term sustainability of Earth systems and services should not stop every effort being made now to improve and maintain current terrestrial targets. However, space activity should be planned with the future support role for the Earth clearly in mind and this will require space activities themselves to be executed in a sustainable way. The view from space, of one world without borders, that is so often the lasting one for astronauts and cosmonauts is one that we should encourage to bring together the necessary global thinking and planning necessary to properly address the world's problems.

4.3.5.2. To industry

It is worth remembering that space is not only of value to industry and commerce as users of services, but also includes the upstream and down-stream supply markets that enable the provision of the services. It is essential for upstream system developers and scientific users to fully understand the environment in which they operate so that they can be guardians of the special resource of the Earth's orbit as well as to be able to plan the protection of the environment beyond Earth's orbit. Government clearly plays a role here within the scope of commitments entered into under space treaties, but it is the industrial sector licensed to use space that must understand the value of the resource and think beyond just today's needs. There are encouraging signs that LEO and GEO telecommunications operators, among other users, have sufficient self-interest and understanding of the issues at hand to work together for the common good and aim to keep the Earth's orbits open for business.

4.3.5.3. To the public

Many will not see space as an enabling factor that provides television and banking services, but do understand and relate to space when human spaceflight is or space exploration – human or robotic – is involved. Many students were encouraged to become involved in mathematics and science due to the early space programmes such as Apollo. Only a selected few were able to become astronauts, but many were inspired by the journey into space and the long-term prospects derived from humans having learnt to live and work in Earth's orbit for longer periods basis.

4.3.6. Conclusions

Space is already helping us to manage our available resources and the human view of Earth from space – one Earth, the blue planet – as a world without borders is an image that should help and inspire us to see the bigger picture. The small band

Fig. 7. *The Blue Marble (source: http://veimages.gsfc.nasa.gov/2429/globe_west_2048.jpg).*[126]

around the Earth's surface which is our fragile atmosphere is hardly visible from a distance, but it is essential to life on Earth as we know it.

4.3.6.1. The Blue Marble

Human spaceflight has been more than just an inspiration, but has been the motivation for much technological development and innovation. Robots will have a long-term role in assisting human exploration and they may be considered the precursors that will provide the insight and understanding that will enable humans to follow their path to the stars.

---

[120] Morison, Ian. "Comet Holmes on 13th November 2007 with Meade 8". Schmidt–Newtonian and Nikon D80 Image. Jordell Bank Observatory. The University of Manchester. 11 Dec. 2007 http://www.jb.man.ac.uk/public/nightsky.html.

[121] During the Apollo missions, many orbital stages were quickly de-orbited to keep the lunar orbits clear for later missions.

[122] "Exploration: NASA's plans to explore the Moon, Mars and Beyond". NASA. 10 Dec. 2007 http://www.nasa.gov/mission_pages/exploration/main/index.html.

[123] COSPAR was founded in 1958 by the International Council for Science (ICSU), at the time called the International Council of Scientific Unions. It is an international organisation charged with the promotion of international collaboration and information exchange in space research. COSPAR. 11 Dec. 2007 http://cosparhq.cnes.fr/.

[124] Two "Sungrazing" comets are seen heading in tandem towards the sun's corona. They do not reappear on the other side. The comets follow similar but not identical orbits and enter the tenuous outer atmosphere of the sun. Shortly after the comets disappeared behind the occulting disks of the coronagraph, a bright helical-shape prominence erupts from the Sun as part of a Coronal Mass Ejections (CME) (explain the acronym). Comets, composed of ice and dust, characteristically have particles streaming out behind them. Comets can be found zooming around space quite frequently. 11 Dec. 2007 http://science.nasa.gov/newhome/pad/images/soho_corona.gif.

[125] Meadows, Donella H., et al. Limits to Growth. New York: Signet, 1972. The book is about the consequences of a rapidly growing world population and finite resources. It was commissioned by the Club of Rome. The five variables in the original model were world population, industrialisation, pollution, food production and resource depletion.

[126] "Visible Earth: a catalog of NASA images and animations of our home planet". NASA. http://veimages.gsfc.nasa.gov/2429/globe_west_2048.jpg.

4.4 Human–machine cooperation in space environments

Gabriella Cortellessa, Amedeo Cesta & Angelo Oddi

4.4.1. Introduction

Space mission environments are characterized by the presence of both human operators and advanced automated technology. One relevant aspect in this context is the degree of interaction between these two entities and, in particular, the role of the human agent with respect to his/her collaboration with *potentially overwhelming* technology. In designing innovative work environments, a certain degree of freedom must be maintained to allow humans and machines to cooperate and adapt to unforeseen contingencies. This paper describes a human–machine cooperation approach to address some of the new challenges introduced by user–system interaction in space missions. Specifically, we will elaborate on the need of retaining a level of flexibility in subdividing responsibilities between autonomous systems and human operators by encouraging the development of *mixed systems* that integrate the capabilities of both entities. Based on our experience in developing frameworks for space missions, we briefly report on two examples of decision-support tools, pointing out the human aspects that need to be taken into account as well as the beneficial effects of synergies between technology developers and experts in different fields like Cognitive Psychology and Human Computer Interaction. While the described experiences are mainly related to space mission control centres, the detected problems as well as the proposed solutions are, to some extent, extensible to manned missions in outer space.

Human space missions have always attracted interest due to the fascinating possibilities to explore new worlds, but also for the new stimulating challenges they present. At present, the main manned long-term experiment is the International Space Station (ISS)[127] but, in the future, manned missions are expected to explore other planets in particular Mars[128] whose distance from Earth is close enough to encourage space agencies into planning to reach it. Human missions in outer space naturally entail new problems and questions that need to be answered. In our vision, humans' activities in space environments involve at least the following aspects:

intelligent technology and humans in spatial contexts should be in the form of *collaboration* rather than master–slave interaction or fully technology-driven work. Since humans and machines need to collaborate, particular efforts are needed to design effective interaction modalities and to create transparent and easy-to-use intelligent systems. The remainder of this paper presents two previous experiences based on this collaborative approach. Even if the intelligent systems developed are devoted to ground segment activities or unmanned space missions, they provide interesting indications for future manned missions.

4.4.3. Lesson learned from experience: two case studies

Two projects we have been involved in allowing us to comment on the validity of the human machine collaboration approach. Both of them synthesize software artifacts that, at different levels, propose a *human users/intelligent system* interaction based on a distribution of active roles. The first project is related to the use of a robotic arm devoted to payload servicing outside the International Space Station, while the second is related to the resolution of the problem of dumping the on-board memory of the Mars Express spacecraft launched by the European Space Agency in 2003 to perform scientific experiments and observations of Mars.

4.4.3.1. SACSO: SAfety Critical SOftware for planning in space robotics

The context of the SACSO project is the robotic arm developed by the Italian Space Agency (ASI) named SPIDER (see Figure 8).

In the SPIDER daily work environment there were two kinds of users:

- The *scientists* responsible for designing and performing the scientific experiments the robot contributes to;
- The *technicians* responsible for controlling the movements of the robot.

The former kind of user was used to reason at a very high level of abstraction and was not aware of or confident with the low level details and technical aspects of the robot. By contrast, the technicians were used to working at a very low level of abstraction, since they were in charge of producing the sequences of commands to operate the robotic arm. For this reason, the interaction between these two groups of people was extremely difficult and ineffective from the point of view of sharing real competences and creating a work environment compatible with a long-term

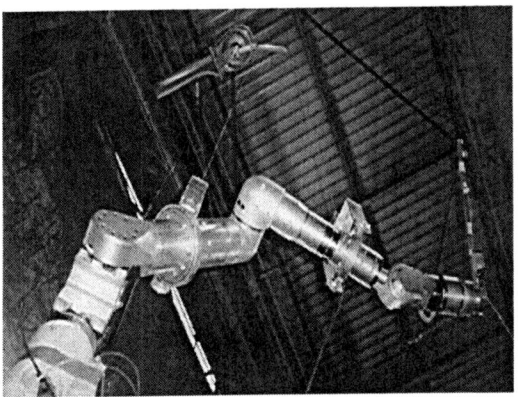

Fig. 8. *SPIDER, the robotic arm developed by the Italian Space Agency (courtesy of ASI) (sources: Italian Space Agency).*

collaboration. In this light, the SACSO project explored the synthesis of a facility in which, *expert users* (e.g., scientists responsible for the scientific experiments) and *technicians* (e.g., robot operators, computer programmers) would cooperate to specify goals and constraints for a robotic arm by means of a high-level specification language and the facility to synthesize the actual robot program. The result of this work was a tool named JERRY[132] which is responsible for the interaction among the different users and an Artificial Automated Planner as well as for the integration of the different modules that compose the overall system, including the management of the data exchange protocols. In particular, JERRY is a modular system for the interactive design, planning, control and supervision of the operation of autonomous robot systems in space. It allows scientists to define the high level objective supported by an environment for intelligent interaction. The problem is then given as input to an Artificial Intelligence (AI) planner. The computed plan is tested by a software simulator, which allows the identification of problems before sending the commands to the real robotic arm on board the space station. These operations can be repeated so as to validate the plan produced by the planner. Once the plan has been validated and tested, an internal translator produces the sequences of instructions to operate the arm, which are directly uploaded to the Space Station (see Figure 9).

The simulation and validation of rover behavior are indeed critical capabilities for scientists and rover operators, to construct, test and validate plan for commanding the robots.[133]

The key AI feature of the project lies in the definition of an experiment as a planning problem, which is then processed by the Planning and Execution modules. The Planning Module requests a high-level description of the task to

139

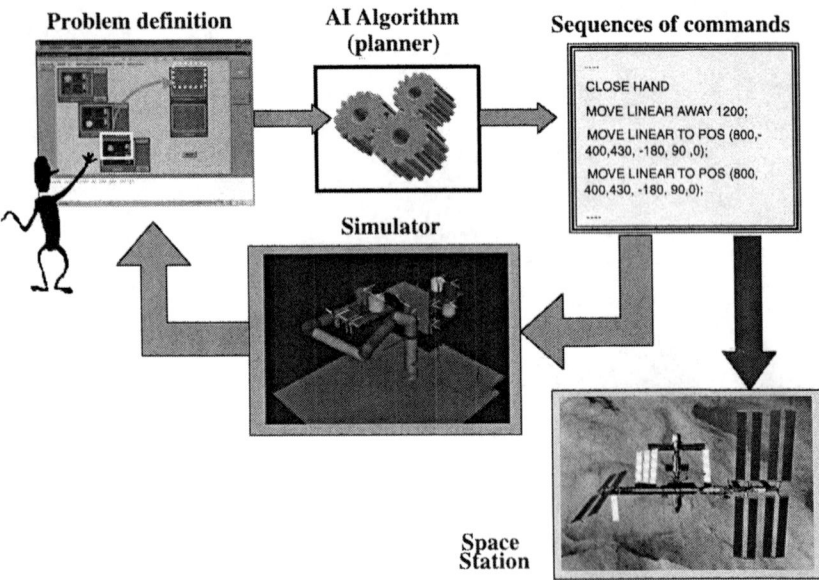

Fig. 9. *Plan synthesis with JERRY. The scientist defines the high level objective with the support of an environment for intelligent interaction. The problem is given to the AI planner, which produces the plan to operate the robot. The plan can be tested and validated by the software simulator. Once the plan has been tested, an internal translator produces the sequences of commands to be sent to the robotic arm on board the space station (sources: ISTC-CNR).*

be performed by the robot and performs the synthesis of an equivalent abstract plan (in a user-oriented symbolic language), i.e., a sequence of high-level actions for the robotic system to execute. The Execution Module transforms an abstract plan (describing the given task as a sequence of high-level actions) into an executable plan where abstract actions are expressed in terms of the basic actions the robotic system can perform, taking into account constraints related to the geometry and physics of the robot system and its operating domain. The executable plan is then encoded into the robot's control language, thus generating an executable code in system-specific language. The resulting code can be visually validated by a software simulator of the robotic system.

In such a highly critical environment, JERRY can effectively support the robot operators in both ordinary and emergency situations and make their work easier, safer and faster. JERRY can also provide scientists with no specific competence in robotics with a higher-level support for the automated execution of complex robot activities, with limited contributions from specialized operators.

This experience allows for a number of conclusions. In particular two main contributions are worth mentioning, which relate to the need of developing trustful and easy-to-use technology and the need to preserve humans' responsibility on critical decisions.

Developing usable and transparent technology. One first comment on the SACSO experience is related to the general difficulty that people usually experience with advanced space technology. This kind of technology is conceived to be used by very specialized users, while on the contrary, scientists are not necessarily technology experts but rather interested in a variety of experiments connected to life in space. In this light, JERRY is intended to provide its functionality to different kinds of users who have to design, control and monitor a robotic arm performing complex tasks, such as the setting up of experiments in a space work cell. The high level goal of the system is to simplify the interaction of users at various levels of expertise with a rather complex robotic device. This consideration highlights the possibility to further investigate innovative representations for space problems as well as new interaction modules within the space contexts, which can bridge the gap between technology and users' expertise.

Preserving user control over technology. A second consideration is related to the need of preserving user control over technology. For this purpose, JERRY has been designed to enable robot operation in Interactive Autonomy, i.e., the system can perform all of its tasks autonomously (including recovery from various non-nominal situations), but the user is able to easily monitor and possibly override autonomous operations, in a collaborative fashion. Effectiveness is guaranteed by a set of tightly integrated specialized modules, each dedicated to a specific task. Interactive Autonomy is attained through a user-centred architecture, where the user asks for services from the specialized modules.

4.4.3.2. The MEXAR2 project

A second project produced a fielded AI system that has been in daily use at the European Space Agency's space operations centre (ESA-ESOC)[134] since February 2005. The tool, named MEXAR2,[135] provides continuous support to human mission planners in synthesizing plans for down-linking on-board memory data from the Mars Express spacecraft to Earth. The introduction of the tool to the mission-planning workflow significantly decreased the time spent in producing plans. Moreover MEXAR2 improves the quality of the produced plans thus guaranteeing a strong reliability in data return and enabling more intense science activity on board. The introduction of MEXAR2 has modified the role of the human mission planners who can now evaluate and compare different solutions rather than dedicate their time exclusively to computing single solutions (a tedious and repetitive task which does not capitalize on the mission planners' decision-making expertise). These characteristics have effectively made MEXAR2 a fundamental work companion for the human mission planners.

The context. A critical problem for interplanetary space missions is maximizing science, while guaranteeing data return to Earth. Additionally, the reduction of investments in the field requires space programs to perform ambitious tasks with more limited budgets with respect to the past. Europe's Mars Express is seen, quoting the mission's web description, "as a pilot project for new methods of funding and working". Specifically, an innovative working relationship between ESA, industry, national agencies and the scientific community as well as the reuse of equipment developed for the ESA Rosetta mission were important factors that have contributed to the development of Mars Express as a relatively low-cost mission. Notwithstanding the low budget, the mission reveals ambitious goals for the scientific experiments on board. The seven payloads the orbiter is equipped with are expected to maximize their data return to take advantage of the opportunity offered by the proximity to the Red Planet. Indeed, an amount of novel information from Mars is arriving to the space science community, and, through the media, to citizens. We may remember the accurate pictures (see examples in Figure 10) taken by the High Resolution Stereo Camera (HRSC) with images of the entire planet in full colour 3D and with a resolution of about 10 m, or the information about the distribution of water, both liquid and solid, in the upper portion of the crust of Mars, distributed by MARSIS, the Mars Advanced Radar for Subsurface and Ionosphere Sounding.

Obviously, in a deep-space mission like Mars Express, data transmission to Earth represents the fundamental aspect. The space probe continuously produces a large amount of data resulting from the activities of its payloads and from on-board device monitoring and verification tasks (the so-called housekeeping data). All these data are to be transferred to Earth during bounded downlink sessions. Moreover, in the case of Mars Express, a single pointing system is present. This implies that, during regular operations, the space-probe either points to Mars to perform payload operations or points to Earth to download the data produced. As a consequence, on-board data are generally required to be stored first in a Solid State Mass Memory (SSMM) and then be transferred to Earth.

The main problem to be solved consists in synthesizing the sequences of spacecraft operations (dump plans) that are necessary to deliver the content of the on-board memory during the available downlink windows. The average data production on a single day can be around 2–3 Gbit while the transmission rate of the communication channel, which varies between 45 kbs and 182 kbs, may not be sufficient. The on-board memory is subdivided into different banks, called packet stores, in which both scientific (science from payloads) and spacecraft management data (housekeeping) can be stored. In particular, housekeeping data

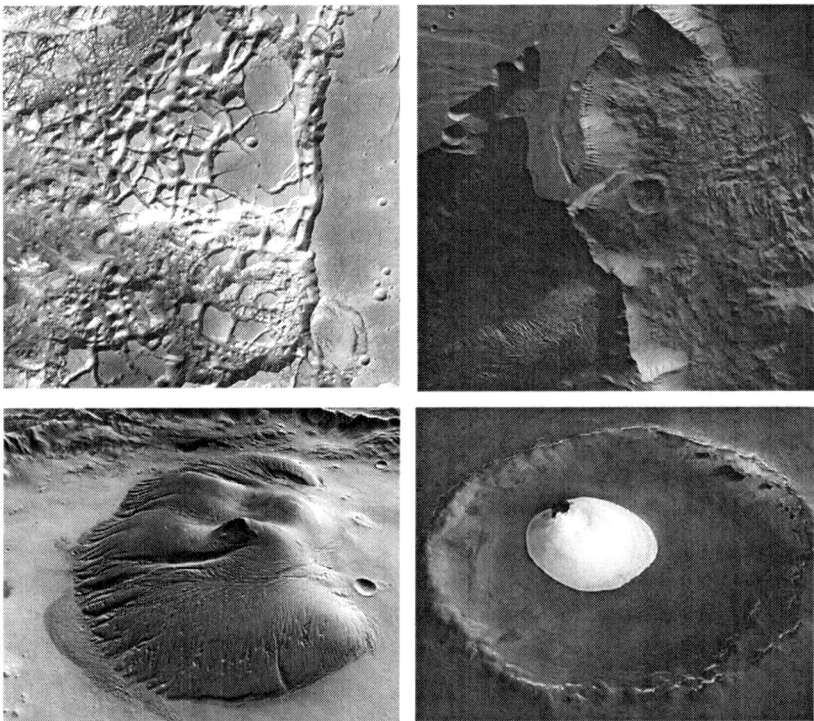

Fig. 10. *Examples of images taken by the High Resolution Stereo Camera on board MARS EXPRESS (sources: ESA).*

must be guaranteed to arrive to Earth daily so as to allow checking for the safety of the different operations on board. Note that each packet store assigned to science data is managed cyclically, so when new data are produced before the previous data are dumped to Earth, the older data are overwritten and the related observation experiments have to be re-scheduled. Even though the on-board memory is about 9.4 Gbit, the irregular distribution of transmission windows, the different transmission rates of such windows and the different data rates for data production (e.g., the stereo camera can produce files close to 1 Gbit) may often create usage peaks close to the packet store capacities. To complicate matters, there is an additional uncertainty factor in data production for some instruments due to different compression algorithms. Usually, dump plans for the on-board memory are computed for a nominal expected production of a certain payload activity, a Payload Operation Request (POR), but mission planners may discover from housekeeping checks that the data on-board are more than expected so they have to recompute the dump plan, i.e., the sequence of dump commands that implements the memory download policy.

The MEXAR2 solution. Given the goal of developing a decision tool for supporting the human mission planner, we have chosen to design a software architecture that captures the entire problem *life cycle*. Specifically, the tool supports a user in all the steps, which run from the definition of an instance of a memory dumping problem to the generation of solutions and their refinement. In previous work practice at ESA-ESOC, the user (mission planner) and the spacecraft (Mars Express) interact through certain modalities. One of the goals of our study has been to contribute an additional means to this interaction by offering a tool that fully preserves the "traditional" real world practice and potentially provides new aids. Indeed, the assumption was made that all the interfaces as defined for the "semi-manual" dump generation tools previously used at European Space Agency (ESA) should be maintained in order to guarantee a smooth transition to the new system. Nonetheless, a goal was pursued to provide a more advanced interactive system that relieves the human planner from boring and repetitive tasks while allowing her to concentrate on more strategic decisions. Our general approach consists in adding a path from the user to the controlled spacecraft. This enhanced path is created through a bipartite architecture composed of a Problem Solver and an Interaction Module. The two modules have distinct roles:

- *Problem Solver.* This module is responsible for modeling the problem and the domain. Based on a CSP (Constraint Satisfaction Problem)[136] approach for the modeling phase, it provides a hybrid solving procedure that combines back-tracking with a maximum flow algorithm to solve problem instances.
- *Interaction Module.* This component is responsible for the dialogue with the user. It supports users in understanding what the solver is doing, improving her trust in the automated solving activity and providing various levels of intervention for strategic decisions during problem solving.

To sum up, in our work we pursued the idea that a user is part of the real world and MEXAR2 endows her with an additional *lens* to analyse the world and act on it. Given this bi-modular framework, the user can concentrate on strategic high level decisions and *what-if* analysis, delegating to the system repetitive and difficult computations. It is clear how the MEXAR2 system has been designed based on the idea of the human–machine cooperation approach. This aspect represented a crucial factor to foster users' acceptance of the tool. A person responsible for a very demanding and critical decision previously performed this task through a semi-manual procedure. MEXAR2, with is collaborative approach, relieved the human mission planner from the hard low-level work by creating a collaborative working environment that empowered human mission planners with additional capabilities.

Again, the human–machine cooperation approach allows exploiting the complementary strengths of humans and machines: precisions and computational power of the artificial solver and creativity and strategic abilities of humans.

4.4.4. Findings and conclusions

Future space missions will be characterized by the presence of humans living for long periods of time surrounded by advanced and very sophisticated technology. Apart from the numerous challenges related to the social human factors that future manned missions pose, an interesting aspect concerns the *role of humans* with respect to their collaboration with *potentially overwhelming* technology and the consequent problem of deciding the *degree of interaction* between humans and machines/robots. This paper has presented examples that contribute to the recent debate in space environments, which propose joint work between humans and machine in space missions. Indeed, these two entities are complementary and equally needed: without autonomous systems, crew members would spend most of their time just trying to stay alive. On the other hand, even with completely autonomous systems crew members would presumably be frustrated by how to repair them, how to ensure they do the right job and how these systems will respond to unexpected events. Human–machine cooperation represents a good compromise to address these problems since it allows exploiting the complementary nature of human and automated reasoning.

These two entities can introduce complementary problem-solving strengths that can be synergistically blended. Often the scale and complexity of practical domains overwhelms the solving capabilities of automated technologies. Likewise, human planners often have additional reasoning resources, which can provide useful strategic guidance, but they are hampered by the complexity of grinding out detailed solutions. In such cases, successful technology application can be achieved through an effective integration of user and system decision-making.

In future manned missions, autonomous systems will need to operate safely in the presence of people and cooperate with them effectively. In this light, we need to design *human-centred* autonomous systems in contrast to traditional *black-box* autonomous ones. Rather than being completely commanded by the users (in a *master–slave* interaction style), or being completely autonomous (*technology-driven* interaction style), these systems enable users to interact with them at different levels of autonomy by tuning the level of control according to needs and preferences (hence *adjustable autonomy*).

145

4.5 Space law in the age of the International Space Station

Frans G. von der Dunk

4.5.1. Introduction

This article focuses on the special context where humans from various nations work and live together in one orbiting laboratory, the International Space Station (ISS), and the legal rules pertinent to those activities. This essentially concerns the application of an existing body of international treaties on space and space activities to the ISS, as well as the special legal framework that has been established to deal with the various ramifications of this very international operating environment. Within that context moreover, the specific European parameters stemming from the fact that the European Space Agency (ESA) serves as the vehicle for the participation of 11 European states in the ISS deserve special attention. The totality of this set of rules, though in several instances not yet elaborated as extensively as might be desired, does provide for a dedicated comprehensive legal framework that may serve as an interesting example of international space law also with a view to future developments.

4.5.2. Towards an International Space Station

4.5.2.1. The background of the initiative to build an International Space Station

Between the moon landings of the early 1970s and the sudden appearance of the prospects of space tourism a few years ago, the most interesting space-related activities were the efforts to build an International Space Station. Although it never captured the imagination of the general public like the Apollo programme did, or even tickled the imagination of some parts of the general public like the sight of adventurous millionaires going into outer space for fun did, the gradual extension of human presence into outer space – as regards duration and scope of activity – through a space station built, launched and operated as a joint international enterprise, was audacious in many ways, not least of all legally.

The idea of launching an international space station evolved from the interest of the U.S. in cooperating with some of its major political partners in the peaceful exploration and use of outer space in a more substantive and consistent fashion than up to then. There had already been a number of essentially bilateral cooperation projects in the area of outer space between the U.S. and, for example, a number of European states collectively such as the Skylab and Spacelab missions. But all of them had been essentially short-term projects, rather than long-term programmes, and none had the same geo-political drive behind them. Thus, in 1984, U.S. President Ronald Reagan mandated the U.S. agency responsible for national civilian space efforts, the National Aeronautical and Space Administration (NASA), to develop a space station and invite relevant partners to join the effort technically, operationally and financially.[137]

4.5.2.2. From the first to the second Intergovernmental Agreement

These efforts led to the first Intergovernmental Agreement of 1988[138] between the U.S., Japan, Canada and a number of European states (ultimately amounting to eleven)[139] represented collectively by the European Space Agency (ESA)[140] on the design, development, operation and utilisation of a space station, which at the time was called "Freedom". The first part of the space station, which was to be assembled in space following a whole series of launches, measured 110 m across and 90 m long with a total weight of about 1 million pounds and was would actually launched in November 1998, with final completion scheduled for 2010 or shortly thereafter.[141]

However, even before the 1988 Intergovernmental Agreement had formally entered into force, the Soviet Union and the attendant political Communist structures fell apart, creating a completely new geo-political paradigm as a backdrop to the whole space station project.

On the one hand, the end of the Cold War meant – for all practical purposes – the removal of political and ideological barriers against using Soviet/Russian technological experience, software and hardware (which in terms of long-duration human spaceflight was by far outstanding versus all the West was able to muster). On the other hand, from the Western perspective, the risk of highly qualified Russian engineers fleeing the financially deteriorating situation at home (where the space industry was no longer a top priority) and seeking employment with whoever was willing to pay was not to be taken lightly. As a result, Russia was successfully invited to join the international partnership. The 1988 Intergovernmental Agreement was renegotiated and ultimately transformed into the 1998 version.[142] "Freedom" was simply renamed "the International Space Station", or "ISS".

For the sake of completeness, it should be added that one more state, Brazil, has effectively become a formal part of the legal construction supporting the ISS venture since then, but as a special partner – namely through a bilateral agreement with the U.S. under the arrangements pertinent to the planned utilisation by Brazil of the U.S. modules of the ISS. In this context, the U.S. had to notify in advance, and seek timely consensus from, the other space station partners.[143]

4.5.2.3. The legal construction underpinning the International Space Station

Overall, the legal construction underpinning the ISS consisted of several layers, with the Intergovernmental Agreement obviously acting as the overarching umbrella for all legal aspects. At a second level, Memoranda of Understanding were concluded between NASA, on the one hand, and the other Cooperating Agencies on the other, to deal with many of the more practical details of developing the ISS. One level further below implementing arrangements were to be concluded whenever necessary between the cooperating agencies concerned.[144]

All the contracts and subcontracts further down the chain, principally between the cooperating agencies and industrial partners charged to develop certain parts of the ISS, were not officially referred to in the Intergovernmental Agreement, yet fall clearly into its scope of application, as well as of the relevant Memoranda of Understanding and implementing arrangements.

A final remark concerns access to the space station. With the accession of Russia to the undertaking of the ISS, transportation to and from the ISS as regards astronauts and cosmonauts was to be offered by Space Shuttle and Soyuz vehicles, while the European partners and Japan were bent on developing cargo vehicles. As a consequence of the basic 'no exchange of funds' philosophy underlying the ISS undertaking, the provision of such transportation services – which for several reasons could not be dealt with feasibly on the basis of 'in kind' compensation – needed to be carved out from that approach by means of a special exception.[145] The 'no exchange of funds' basis was not applied to Russia in view of the aforementioned rationale for taking Russia on board, but this was also clearly an exception dictated by ulterior motives, without impinging upon the underlying philosophy of the joint venture.

4.5.3. The novelty of the International Space Station

4.5.3.1. From short-haul flights to a long-haul presence in outer space

Politically, of course, the inclusion of Russia, the former Cold War enemy of the other partner states in the cooperative venture, which was shaped in the 1990 and culminated in the 1998 Intergovernmental Agreement, was already a novelty – at least for such a highly visible, high tech-area with many security-sensitive aspects as building a station orbiting in outer space.

More important from a legal perspective at least was the envisaged quasi-permanency of the station. While the longevity of MIR, resulting in nearly 15 years of orbital lifetime, was in many respects a matter of surprise – as well as keeping it alive on a shoestring almost literally – the ISS was from the start destined to serve for decades, as "a permanently inhabited civil international space station."[146] Hence, from the start, it was also envisaged to serve a wider variety of human activities, far beyond the mere traffic or station-keeping activities thus far key to any legal concerns. It meant that various legal regimes other than space law properly speaking could, would or should now become applicable to those activities as well, varying from criminal law to intellectual property rights relating to the protection of inventions made on board the ISS.

4.5.3.2. The international character of the ISS venture and the position of Europe

The unique international character of the whole undertaking came to be duly reflected in the legal construction. Legally speaking, all the preceding space stations constituted simple legal constructs as single-nation stations even if many foreign crew visited them. Following Article VIII of the Outer Space Treaty[147] and Article II of the Registration Convention,[148] the stations were registered by the respective states, and thereby qualified as their quasi-territory for legal purposes. Likewise, for example, liability for damages caused by the operations of such space stations would revert to those states in accordance with Article VII of the Outer Space Treaty and Articles I–V of the Liability Convention.[149]

With the refusal by the other partner states already under the 1988 Intergovernmental Agreement to simply register the whole ISS as a U.S. space object, the involvement of many jurisdictions came into play. Thus, under the Intergovernmental Agreements (both the 1988 and the 1998 versions) "each Partner shall register as space objects the flight elements (…) which it provides", and consequently will be principally entitled to "retain jurisdiction and control" over such elements as well as the personnel on board.[150] In other words: legally speaking, the ISS consisted of a number of floating pieces of quasi-territory of the different states (Canada effectively being excluded as it was not to provide any manned element) linked to each other in the global commons of outer space.

A further unique element in this context was the explicit designation of ESA as a partner in the ISS undertaking.[151] ESA, as an intergovernmental organisation consisting of sovereign member states, does not and cannot exercise jurisdiction and legal control in the normal sense of the word.[152] However, under certain conditions, the Registration Convention allows intergovernmental organisations to serve as the equivalent of a state for all practical purposes regarding the legal regime vis-à-vis parties – ESA indeed complies with those requirements and can thus effectively act as registration 'state' for the European module of the ISS.[153] This means that for any legal issues requiring the exercise of such 'real' jurisdiction and legal control, reference will (have to) be made to the individual member states of ESA participating in the ISS as parties to the 1998 Intergovernmental Agreement.

4.5.3.3. Commercialisation of ISS utilisation and 'space tourism'

Though originally not taken into consideration,[154] it soon became clear that another novelty given birth by the ISS would be commercial utilisation. It was partly the continued problems with governmental funding within various partner states that led them to start considering, in the late nineties, the possibilities of generating interest – and investments – among private companies in using the ISS as opposed to merely being subcontracted to build elements or to use certain space-based products or services derived from ISS activities. The microgravity environment was considered of great potential interest, in particular, for medical and chemical industries, but other semi- or proto-commercial uses were also expected.

Thus, ESA, for example, was given the mandate in 1999 to promote the commercial utilisation of the European module of the ISS, and officially such usage up to 33% in terms of available time was envisaged.[155] Other partners arrived at similar constructions internally.[156] This partial commercialisation resulted once more in a broader scope of legal issues being at least potentially applicable to ISS operations and bringing in existing regimes such as liability and contract-related law – or even necessitating new legal instruments such as an ISS Crew Code of Conduct and a Multilateral Crew Operations Panel, all under the umbrella of the Intergovernmental Agreement.[157]

The most spectacular novelty, certainly for outsiders, was of course the advent of 'space tourism', which took its first aim at the ISS. In April 2001, U.S. citizen Dennis Tito was launched to the Russian part of the ISS for no other reasons than that he was driven by his desire to fly in outer space and happened to have the money privately available to pay the price quoted to him for fulfilling that desire.[158]

Originally Tito, through the brokerage of a small private company called MirCorp established specifically for bringing self-financed private persons into space, was supposed to be launched on a Russian launch vehicle to the Russian space station Mir, at an overall price estimated at the time to amount to some USD 20 million. In the course of his preparation, however, Mir had to be de-orbited, which occurred over the Pacific Ocean in March 2001.[159] In order to honour their contractual commitment, the Russians had only one way out: to change Tito's destination to the Russian module of the ISS which was being built at the time.

Paying similar amounts for the privilege, a second millionaire, South-African Mark Shuttleworth, followed suit in 2002; U.S. national Greg Olsen became the third space tourist in 2005; and Anousheh Ansari became the first female space tourist in 2006 – and all had the ISS as the destination for their week-long stay in outer space.[160] It also became clear that this new branch of space activities created the distinct necessity to establish appropriate legal rules and principles, and to some extent – essentially at the national level within the U.S. – this has already been done.[161]

4.5.4. Space law and the International Space Station

Some of the major elements of the application of international space law to the ISS were briefly discussed in the preceding part, with a view to some novel characteristics as these have then been addressed in the Intergovernmental Agreement, but they merit a second look now.

4.5.4.1. Jurisdiction in general

Thus, the jurisdiction of individual partner states was seen to apply to respective parts of the ISS through the registration of the separate elements of the ISS as separate space objects, which as such was in compliance with the framework regime offered by Article VIII of the Outer Space Treaty and relevant clauses of the Registration Convention.[162] However, how any potential conflict of laws was to be dealt with, for example when a U.S. and a Japanese astronaut/engineer would be involved in a legal issue on board the Russian module, was not elaborated further by the Intergovernmental Agreement, with two prominent exceptions. These concerned specific issues of jurisdiction considered immediate and important enough that they should not be permitted to be dealt with only by the time a conflict would have arisen and/or by means of general principles of conflict of laws: criminal jurisdiction and intellectual property rights jurisdiction.

4.5.4.2. Criminal jurisdiction

As for criminal jurisdiction,[163] the prospect of the application of criminal laws on board of the ISS in view of the long duration of human presence in a limited space with a limited amount of others coming from various cultures and backgrounds was considered substantial enough, in spite of the extensive screening and training of astronauts and cosmonauts (and in spite of the cooperative approach to the whole venture) to warrant more detailed arrangements already at the level of the IGA itself.

Interestingly, dealing with this issue was to present one of the few key differences between the 1988 and 1998 versions of the Intergovernmental Agreements. In the older version, the case of a certain activity or event involving someone present on board the ISS raising questions of potential criminal liability was dealt with by application of quasi-territorial jurisdiction. The jurisdiction of the state on board of whose element that activity or event had occurred would apply in first instance; however, U.S. authorities could exercise their jurisdiction potentially overriding any other one if the activity or event posed a threat to the safety of operations on board the ISS[164] – which, as one can imagine, could quite easily be the case.

In fact, this potentially overriding U.S. jurisdiction reflected the fact that any person to be prosecuted for acts on board of the ISS, under the old construct, could only be brought back to earth by means of a U.S. vehicle, as no other partner state at the time possessed manned spaceflight capability. Hence U.S. jurisdiction and control would have first choice.[165] Obviously, that changed once Russia came on board, and since Russia did not appreciate this construct the relevant clauses were altered in the course of the negotiations.

Thus, Article 22(1) of the 1998 Intergovernmental Agreement now reads: "Canada, the European Partner States, Japan, Russia, and the U.S. may exercise criminal jurisdiction over personnel in or on any flight element who are their respective nationals." This is in essence the so-called 'active personality-principle', well-known in general international law as a justification for exercising criminal jurisdiction.[166] Article 22(2) then adds certain possibilities for other states to exercise their jurisdiction on the basis of passive nationality[167] or quasi-territoriality,[168] but this depends on the extent to which the state of nationality of the alleged perpetrator itself is interested in prosecution.

4.5.4.3. Jurisdiction and intellectual property rights protection

While the application of criminal jurisdiction circumvented the additional problem of ESA-involvement, read ESA-registration of the European module, by applying the active personality-principle – ESA astronauts qualifying as nationals of respective ESA member states, with the qualification of part of the ISS as an 'ESA module' not being relevant – a different approach was taken for jurisdiction relevant for applying intellectual property rights. Article 21(2) of the 1998 Intergovernmental Agreement applies the quasi-territorial approach, in that sense following the general regime of Article 5: "for purposes of intellectual property law, an activity occurring in or on a Space Station flight element shall be deemed to have occurred only in the territory of the Partner State of that element's registry."

Obviously then, in this case a further solution had to be found for the specific European context, where ESA does not have any "territory" in the legal sense of the word. Thus, "for ESA registered elements any European Partner State may deem the activity to have occurred within its territory".[169] To date, two European states – Germany and Italy – have actually taken the trouble to extend the scope of their national, territorially-based legislation protecting inventions by means of patents to inventions done on board of the European module of the ISS.[170] The result is that anyone entitled to claim a patent as regards an invention done on board of the European module of the ISS, whether of German, Italian or any other nationality (European or otherwise) should register his or her patent with either the German or the Italian authorities.

Firstly, based on the rather advanced measure of harmonisation of intellectual property rights within Europe, the protection under such a registration does not only extend to those other European states but is also of a similar scope and nature.[171] Secondly, based on conventions going back as far as 1883 (the so-called Paris Convention)[172] and the activities of the World Intellectual Property Organisation (WIPO),[173] such patents would turn out to be basically protected in most jurisdictions across the globe.

4.5.4.4. Jurisdiction and 'space tourists'

A further issue, related to some extent to the general one of jurisdiction, arose suddenly some years after the Intergovernmental Agreement: the visit of the world's first space tourist in 2001 triggered, among other things, a discussion regarding the terminology used in the space treaties of "astronauts"[174] and "personnel" of a spacecraft,[175] which entailed certain privileges pertaining to the special obligations of the relevant states to not only come to the rescue in case of distress, but also provide support in returning the persons concerned home as speedily as possible, without entertaining any thoughts about applying domestic jurisdiction, for whatever, reason, to those persons prior to their return home.

The ISS Crew Code of Conduct created another category of "spaceflight participants", where the term 'spaceflight participant' referred to "an individual (e.g., ... crewmembers of non-partner space agencies, engineers, scientists, teachers, journalists, filmmakers or tourists), sponsored by one or more partner(s); normally this is a temporary assignment that is covered under a short-term contract; they are eligible for assignment as visiting scientist, commercial user or tourist, but their task assignment cannot include ISS assembly, operations and maintenance activities".[176] By encompassing space tourists, it ensured that these would *not* enjoy those same privileges as astronauts or cosmonauts.

4.5.4.5. The issue of liability for damage

The next major issue that was dealt with at the level of the Intergovernmental Agreement itself in very fundamental terms, considering the potential threat it constitutes to the general cooperation spirit behind the whole venture, concerned the question about what should happen, should damage occur within the context of any of the activities related to the design, development, operation and utilisation of the ISS.

Space law as it stood under the Liability Convention did provide for a somewhat elaborate system of liability for damage caused by space activities, more precisely caused by space *objects*; such liability was allocated to the launching state(s).[177] Further clauses provided, for example, for the applicability of absolute respective fault liability, for joint and several liability, for the lack of a limit to compensation, for a *jus standi* under the Convention, for exculpatory clauses and for a rudimentary dispute settlement system.[178] As may clearly be derived from many clauses in the Liability Convention, however, the liability system was very much geared to third-party liability, and not very helpful for application to cases of inter-party liability, even though not formally excluded by the Convention.[179]

The Intergovernmental Agreement acknowledges this regime as being applicable to any damage caused by the ISS or any of its elements as space objects to third states,[180] and then creates by means of Article 16 an extended regime for dealing with intra-party damage and the question of liability. Essentially, it provides for a cross-waiver of liability for damage caused in the context of what has been defined in a sweeping manner as "Protected Space Operations",[181] between not only the partners and partner states themselves, but also between any or all of the "related entities" of one such partner state and those of another partner state.[182] There are a few exceptions to this cross-waiver, such as when it concerns private claims for bodily injury or death or claims for damage caused by wilful misconduct,[183] but overall, the spirit of cooperation has resulted in the need for each partner state and their related entities to simply accept the possibility that they may suffer damage in the context of the ISS without being able to assert a liability claim for the purpose.

4.5.5. What comes next?

In sum, the current legal arrangements at the level of the ISS itself, even as regards the main legal document which is the 1998 Intergovernmental Agreement, represent a highly interesting and innovative set of legal rules, rights and obligations resulting from the need to deal with the truly international character of the ISS. In particular, the additional novel element of the role of the European Space Agency as an intergovernmental organisation has called for additional innovations in law-making. The ISS, novel as it may be also in legal terms, represents no more than an intermediate step in the broader adventure of mankind's expansion into outer space. Yet, in many respects, the ISS forebodes the legal issues to follow such expansion – and the legal construct supporting it has already come up with helpful legal solutions in some cases.

This is not the proper place to dwell long on the many, often futuristic, plans for future activities in outer space pertaining to long-duration human presence. It should suffice to say that with the human presence in outer space continuously being extended – and, presumably, made easier and cheaper by a magnitude or two – the number of legal issues that will become relevant at least theoretically will inevitably grow. When mankind actually establishes 'space colonies' on celestial bodies, at least in the non-legal sense of the word because up to now colonisation in the legal sense is clearly prohibited by Article II of the Outer Space Treaty, many of the issues discussed with regard to the ISS legal regime – criminal liability, intellectual property rights, liability for damage will become even more prominent. Moreover, new issues such as the nationality of space-born babies, the applicability of human rights to outer space, and the validity of contracts drawn up in outer space on outer space matters will present themselves in due course.

This may trigger discussions on whether jurisdiction, which is currently not possible on a territorial basis, should be structured differently so as to ensure that law will actually follow man into outer space. Or will jurisdiction based on the nationality of the humans involved suffice – but then, what about these future space-born humans? Even now, with the impending prospects of man returning to the moon and then on to Mars, issues such as safety- and/or security-zones around installations on celestial bodies, the exploitation of mineral resources *in situ* and other commercial issues such as licensing are back on the table. The fate of the Moon Agreement[184] should warn us: after it had been drafted with the involvement and general consent of all important states concerned, a swift change in the global political climate caused all those to renege on actually ratifying and in most cases even signing it. Thus, it currently has only thirteen parties (including none of the major space-faring nations) plus four states only having signed the Agreement (including France and India); its relevance in legal terms is therefore to be severely doubted. Clearly, therefore, there is no easy road when it comes to building a legal regime acceptable and fair to all, as well as workable and efficient – but inevitably it is a road we must take, as the alternative would be considerably worse: a legal near-vacuum in outer space.

---

[137] Rosmalen, S. "The International Space Station Past, Present and Future: An Overview". The International Space Station: Commercial Utilisation from a European Legal Perspective. von der Dunk, Frans G. and Brus, Marcel M. T. A. eds. Leiden: Brill, 2006. pp. 9–14.

[138] Agreement among the Government of the United States of America, Governments of Member States of the European Space Agency, the Government of Japan, and the Government of Canada on Cooperation in the Detailed Design, Development, Operation, and Utilization of the Permanently Manned Civil Space Station (hereafter 1988 Intergovernmental Agreement), Washington, done 29 September 1988, entered into force 30 January 1992. Space Law – Basic Legal Documents, D.II.4.2.

[139] This concerned Belgium, Denmark, France, Germany, Italy, the Netherlands, Norway, Spain, Sweden, Switzerland and the United Kingdom.

[140] ESA was established by means of the Convention for the Establishment of a European Space Agency (hereafter ESA Convention), Paris, done 30 May 1975, entered into force 30 October 1980. 14 ILM 864 (1975). As of this writing, ESA counts seventeen member states. ESA pools the financial and technical resources of its member states to conduct research and development activities *vis-à-vis* or in outer space; the European contributions to the ISS are undertaken as optional programmes, in accordance with Art. V(1.b), ESA Convention, which means *inter alia* that not necessarily all ESA member states have to participate.

[141] Rosmalen, S. "The International Space Station Past, Present and Future: An Overview". The International Space Station: Commercial Utilisation from a European Legal Perspective. von der Dunk, Frans G. and Brus, Marcel M. T. A. eds. Leiden: Brill, 2006. pp. 11–14.

[142] Agreement among the Government of Canada, Governments of Member States of the European Space Agency, the Government of Japan, the Government of the Russian Federation, and the Government of the United States of America concerning Cooperation on the Civil International Space Station (hereafter 1998 Intergovernmental Agreement), Washington, done 29 January 1998, entered into force 27 March 2001. Space Law – Basic Legal Documents, D.II.4.

[143] Art. 9(3.a), 1998 Intergovernmental Agreement.

[144] Art. 4, 1998 Intergovernmental Agreement.

[145] This exception was provided by Art. 12, 1998 Intergovernmental Agreement, which in par. 2 allows for the provision of such services on "a reimbursable (. . .) basis".

[146] Art. 1(1), 1998 Intergovernmental Agreement.

[147] Treaty on Principles Governing the Activities of States in the Exploration and Use of Outer Space, including the Moon and Other Celestial Bodies (hereafter Outer Space Treaty), London/Moscow/

Washington, done 27 January 1967, entered into force 10 October 1967. 610 UNTS 205. TIAS 6347. 18 UST 2410. UKTS 1968 No. 10. Cmnd. 3198. ATS 1967 No. 24. 6 ILM 386 (1967).

[148] Convention on Registration of Objects Launched into Outer Space (hereafter Registration Convention), New York, done 14 January 1975, entered into force 15 September 1976. 1023 UNTS 15. TIAS 8480. 28 UST 695. UKTS 1978 No. 70. Cmnd. 6256. ATS 1986 No. 5. 14 ILM 43 (1975).

[149] Convention on International Liability for Damage Caused by Space Objects (hereafter Liability Convention), London/Moscow/Washington, done 29 March 1972, entered into force 1 September 1972. 961 UNTS 187. TIAS 7762. 24 UST 2389. UKTS 1974 No. 16. Cmnd. 5068. ATS 1975 No. 5. 10 ILM 965 (1971).

[150] Art. 5(1), resp. (2), 1998 Intergovernmental Agreement, which furthermore explicitly refers to Art. VIII, Outer Space Treaty, and Art. II, Registration Convention.

[151] Art. 3(b), 4, 1998 Intergovernmental Agreement.

[152] Shaw, Malcolm N. International Law. 3rd ed. Cambridge: Grotius Publications Limited, 1991. p. 393. "Jurisdiction concerns the power of the state to affect people, property and circumstances and reflects the basic principles of state sovereignty, equality of states and non-interference in domestic affairs." Thus, it is an exclusive prerogative belonging to states.

[153] Art. VII(1), Registration Convention, in conjunction with Art. 5(1), 1998 Intergovernmental Agreement. ESA has formally complied with the requirements posed by the Registration Convention by a Declaration of 2 January 1979. Space Law-Basic Legal Documents, A.IV(4.2). 2.

[154] Art. 9, 1998 Intergovernmental Agreement, on "Utilization" only referred to private entities in a general sense, and/or whether activities concerned would be conducted for peaceful purposes.

[155] Belingheri, M. "A Policy and Legal Framework for Commercial Utilisation". The International Space Station: Commercial Utilisation from a European Legal Perspective. von der Dunk, Frans G. and Brus, Marcel M. T. A. eds. Leiden: Brill, 2006. p. 35.

[156] Veldhuyzen, R. P. and Masson-Zwaan, T. L. "ESA Policy and Impending Legal Framework for Commercial Utilisation of the European Columbus Laboratory Module of the ISS". The International Space Station: Commercial Utilisation from a European Legal Perspective. von der Dunk, Frans G. and Brus, Marcel M. T. A. eds. Leiden: Brill, 2006. p. 51.

[157] Ibid. p. 53.

[158] Launius, Roger D. and Jenkins, Dennis R. "Is it finally time for space tourism?" Astropolitics 4 (2006): 255.
Smith, Lesley J. and Hörl, Kai-Uwe. "Legal Parameters of Space Tourism". Proceedings of the Forty-Sixth Colloquium on the Law of Outer Space. Washington: American Institute of Aeronautics and Astronautics Inc, 2004. p. 38.

[159] Billings, Linda. "Exploration for the masses? Or joy-rides for the ultra-rich? Prospects for space tourism". Space Policy 22 (2006): 163.
Sattler, Rosanna. "U.S. Commercial Activities aboard the International Space Station". Air & Space Law 28 (2003): 81.

[160] Launius, Roger D. and Jenkins, Dennis R. "Is it finally time for space tourism?" Astropolitics 4 (2006): 260.

[161] Most of those are not particularly or exclusively relevant in the context of the space station, if only since the main thrust of 'space tourism' is rapidly shifting to the sub-orbital version currently being operationalised by the likes of Virgin Galactic; hence they will not be dealt with here. For further discussion of this new area of space law reference may soon be had to Von der Dunk, F. G. Passing the Buck to Rogers: International Liability Issues in Private Spaceflight. This article is to be published in the Nebraska Law Review in 2008.

[162] Art. II, Registration Convention. Art. IV furthermore provides the main parameters to be included in the international registry with the UN Secretary-General, who in turn had delegated running the international registry to the UN Office for Outer Space Affairs (OOSA). UNOOSA. http://www.unoosa.org/oosa/en/SORegister/index.html.

[163] "Criminal jurisdiction" is defined as "power of tribunal to hear and dispose of criminal cases", whereas "criminal" is defined as "that which pertains to or is connected with the law of crimes, or the

administration of penal justice, or which relates to or has the character of crime". West's Law & Commercial Dictionary in Five Languages, Vol. I. St. Paul: West Publishing Company, 1985. pp. 386–387.

[164] Art. 22, 1988 Intergovernmental Agreement.

[165] The legal result thereof would be that any person of a different nationality than the U.S. one could only become effectively subject to the jurisdiction of his state of nationality after extradition by the U.S. to his state of nationality, extradition being a well-known international law-phenomenon hinging upon treaties calling for or even merely allowing for relevant cases of extradition.
Shaw, Malcolm N. International Law. 3rd ed. Cambridge: Grotius Publications Limited, 1991. pp. 422–423.
Wallace, Rebecca M. M. International Law. 3rd ed. London: Sweet & Maxwell, 1997. pp. 119–121.

[166] Shaw, Malcolm N. International Law. 3rd ed. Cambridge: Grotius Publications Limited, 1991. p. 403.

[167] Ibid. pp. 408–409.

[168] Ibid. pp. 402–403.

[169] Art. 21(2), 1998 Intergovernmental Agreement.

[170] Balsano, A. M. and Wheeler, J. "The IGA and ESA: Protecting Intellectual Property Rights in the Context of ISS Activities". The International Space Station: Commercial Utilisation from a European Legal Perspective. von der Dunk, Frans G. and Brus, Marcel M. T. A. eds. Leiden: Brill, 2006. pp. 65–67.

[171] This harmonisation process essentially takes place under the guidance of the Convention on the grant of European Patents, Munich, done 5 October 1973, entered into force 7 October 1977. p. 1065 UNTS 199. Cmnd. 7090; the Convention for the European Patent for the common market, Luxembourg, done 15 December 1975. Cmnd. 6553. 15 ILM 5 (1976). OJ L 401/9 (1989); and the Agreement relating to Community patents (89/695/EEC), Luxembourg, done 15 December 1989. OJ L 401/1 (1989).

[172] Convention for the Protection of Industrial Property as Modified by Additional Act of 14 December 1900 and Final Protocol (hereafter Paris Convention), Paris, done 20 March 1883, entered into force 6 July 1884. USTS 379. UKTS 1907 No. 21. ATS 1907 No. 6. Later treaties further added to the scope and extent of global application and protection of patents and related rights, such as the Patent Cooperation Treaty, Washington, done 19 June 1970, entered into force 24 January 1978. 1160 UNTS 231. TIAS 8733. 28 UST 7645. Cmnd. 4530. UKTS 1978 No. 78. ATS 1980 No. 6. 9 ILM 978 (1970).

[173] WIPO was established by means of the Convention Establishing the World Intellectual Property Organisation (WIPO), Stockholm, done 14 July 1967, entered into force 26 April 1970. 828 UNTS 3. TIAS 6932. 21 UST 1749. UKTS 1970 No. 52. Cmnd. 3422. ATS 1972 No. 15. 6 ILM 782 (1967).

[174] Art. V, Outer Space Treaty; title and Preamble of the Agreement on the Rescue of Astronauts, the Return of Astronauts and the Return of Objects Launched into Outer Space (hereafter Rescue Agreement), London/Moscow/Washington, done 22 April 1968, entered into force 3 December 1968. 672 UNTS 119. TIAS 6599. 19 UST 7570. UKTS 1969 No. 56. Cmnd. 3786. ATS 1986 No. 8. 7 ILM 151 (1968).

[175] Art. VIII, Outer Space Treaty; Art. 1, 2, 3, 4, Rescue Agreement.

[176] Veldhuyzen, R. P. and Masson-Zwaan, T. L. "ESA Policy and Impending Legal Framework for Commercial Utilisation of the European Columbus Laboratory Module of the ISS". The International Space Station: Commercial Utilisation from a European Legal Perspective. von der Dunk, Frans G. and Brus, Marcel. M. T. A. eds. Leiden: Brill, 2006. p. 55.

[177] Art. I(c), Liability Convention.

[178] Art. II, III, IV, V, XII, VIII, VI, XIV–XX, Liability Convention.

[179] Such evaluation arises *inter alia* from clauses referring to cases involving more than one state in the causation of damage, where only the inter-party distribution of third-party liability was referred to, which depending upon the case was then explicitly or implicitly left for those states to deal with (Art. V (2), resp. Art. IV(2), Liability Convention), Art. III referring to damage done to the space object of

another state, and Art. VII(b) *inter alia* excluding "foreign nationals (. . .) participating in the launch" from the scope of the Convention in case they suffer damage caused by the space object thus launched.
[180] Art. 17, 1998 Intergovernmental Agreement.

[181] Art. 16(2.f), 1998 Intergovernmental Agreement. This article defines such Protected Space Operations as "all launch vehicle activities, Space Station activities, and payload activities on Earth, in outer space, or in transit between Earth and outer space in implementation of this Agreement, the MOUs, and implementing arrangements", then taking care to even further elaborate the broad scope of the concept with further examples.

[182] Art. 16(1), 1998 Intergovernmental Agreement. Art. 16(2.b) then defines "related entity" again very broadly as "(1) a contractor or subcontractor of a Partner State at any tier; (2) a user or customer of a Partner State at any tier; or (3) a contractor or subcontractor of a user or customer of a Partner State at any tier".

[183] Art. 16(3d), 1998 Intergovernmental Agreement.

[184] Agreement Governing the Activities of States on the Moon and Other Celestial Bodies (hereafter Moon Agreement), New York, done 18 December 1979, entered into force 11 July 1984. 1363 UNTS 3. ATS 1986 No. 14. 18 ILM 1434 (1979).

5.2 Humans – more than the better robots for exploration?

Wolfgang Baumjohann

5.2.1. Introduction

The scientific exploration of outer space, defined here as going to places between the Earth's ionosphere at 100 km altitude and the outer reaches of our solar system, some 10 billion kilometres away, and doing measurements there, has been done mostly by robots. Why? For the outer solar system, beyond the orbit of Mars or the asteroid belt, the answer is simple: we do not yet have the technology to let humans go there, let alone survive there. In the inner solar system, Venus is within reach in terms of travel time, but both Venus and Mercury's conditions are so hostile that the survival of humans would not be possible.

That leaves the Moon and Mars, possibly near-Earth asteroids, and the near-Earth space area in general. For these objects and regions, the answer depends on a subtle difference in the way the question is posed. Shall humans go *to* explore the Moon and Mars? *No.* Shall humans go *and* explore the Moon and Mars? *Yes.*

5.2.2. Scientific exploration

Scientific exploration of outer space begun in 1958 with the launch of the first scientific satellite, Explorer-I, into the Earth's orbit, which led to the discovery of the Van-Allen belts (Sputnik-I was launched 4 months earlier, but did not carry any instruments). While Van Allen's Geiger counters can hardly be called robots (and indeed the first scientific measurement in space was actually a non-measurement. The Geiger counters became saturated because the energetic particle flux was much higher than expected), and with the advent of the computer age, the scientific instruments and space probes became more and more sophisticated and now can rightfully be called robots. In particular the Mars Exploration Rovers Spirit and Opportunity, which have been roving Martian soil since early 2004, are role models of robots even by their appearance.

In situ measurements done by robots have enlarged our knowledge exponentially about Earth's space neighbourhood and our solar system. Most planets and a number of minor bodies have been visited at least during fly-bys. More orbiters and landers are underway, are being built, or are at least in the planning stage.[185] The question is: would humans have done better? Yes, of course, and the Apollo programme has shown that humans are still better at science than robots. But it comes at a price, a price too high to pay for science.

Already at the inner edge of space, in low-Earth orbit, having humans do experiments is considerably more expensive letting robots do them on an automated re-entry module like Foton-M. Even biological and medical research can be done during such a relatively cheap unmanned mission.[186] The difference in price becomes much larger if one compares robotic sample return missions to establishing a lunar base or a return trip to Mars. One hundred billion and 400 billion Euros have been estimated for establishing a lunar base and a combined Moon–Mars programme,[187] respectively. Comparing these numbers with the cost of a robotic rover, roughly one billion euros, or a sample return mission at, say two to three billion euros, one might argue that the potential frequency of robotic missions might overcome the ability of humans to respond to unforeseen circumstances.

However, the 13-digit figures in euro or U.S. dollars (or even more digits in yuan) are most likely not the only price to be paid. Unfortunately, on several occasions we have seen that human spaceflight can cost a priceless quantity: human life. In the past, these disasters were caused by failures of human technology. In the future, thinking about lunar exploration or going to Mars, another equally dangerous threat comes into play: stormy space weather.[188]

The sun continuously emits a stream of energetic particles, the so-called solar wind. At times big eruptions in the sun's outer layer, so-called coronal mass ejections, catapult huge blobs containing even more of these energetic ions and electrons. When such blobs of solar plasma hit the Earth, most of the potentially hazardous particles are deflected by the Earth's magnetic field, which extends tens to hundreds of thousand kilometres into outer space and serves as a magnetic barrier that shields the so-called magnetosphere,[189] at least partially, from these dangerous events. Even so, a number of (automated) spacecrafts have stopped functioning during such events, most likely due to severe damage of their electronic components by electrons and ions with energies of some hundred thousand to millions of electronvolts.

Thus far, no human life has been lost in these space storms, mainly because humans hardly left low-Earth orbit, where the magnetic shield is still very strong and deflects most of the damaging particles. The situation is different when travelling to the Moon. At lunar orbit, the terrestrial magnetic field is already quite weak and cannot serve as a protective shield anymore. In fact, several of the Apollo missions missed space storms by days or weeks only whose millions of electronvolts protons would have caused a high cancer risk or severe radiation sickness for the astronauts and a potential 'crew-killing' event occurred between the Apollo 16 and 17 missions.[190]

There is still hope of overcoming this problem for lunar missions. The coronal mass ejections and their hazardous particles need 3 to 4 days to travel from the sun to the Earth, just about the same time astronauts need to travel from the Earth to the Moon. With the advent of modern solar observatories like the recently launched Stereo pair of spacecraft,[191] one is able to see the birth of coronal mass ejections, allowing the delay of a launch until the storm subsides.

For astronauts, cosmonauts, or taikonauts, on their months-long journey to Mars (even assuming that not yet-available nuclear propulsion will allow shorter transits), the situation is different. Their spacecraft will be hit by the full force of the space storms and they could die or at least suffer severe radiation damage unless their spacecraft is heavily shielded. Shielding is, in principle, possible, but comes at the cost of adding substantial weight for thick layers of metals or composites, water tanks, or devices generating powerful magnetic shields. Developing a light-weight way of shielding against solar storms and the even more hazardous cosmic rays, which originate from supernova explosions, will be essential to make interplanetary travel different from sitting close to a leaky nuclear reactor and to make the travel of humans to Mars (and beyond) possible at all.

5.2.3. Real exploration

Real exploration in the classical sense, i.e., going to new frontiers and expanding the human sphere into outer space, is a different issue. In this context, the cost is not as much an issue, and also the risk of life is not out of question anymore. In fact, while people risking their life for gaining new scientific insight are likely to be labelled suicidal, people dying for whatever greater goal become heroes. Like in the exploration of the polar regions, the first ascents of the 8000-m peaks in the Himalaya, as well as in the Apollo mission, national prestige and patriotism come into play and render the concept of cost, be it financial or possible loss-of-life, secondary. Even many individuals who are not advocates of patriotism or national pride believe that real exploration is worth a higher cost envelope than purely scientific exploration (including the author of this article). While most space scientists will not agree that home-delivery of a few kilograms of Martian stones by human hands is worth half a trillion euros and the potential hazard to human life, deep in their heart they feel like all human beings: exploration and striving for new frontiers are an integral part of humanity. In real exploration, robots can do reconnaissance, but cannot replace humans.

Like many real explorers, the space explorers will most certainly do scientific work during their journeys as well. Good examples are the polar explorers. Fridtjof Nansen's first priority certainly was not to gain new scientific insights on the polar region, but his two-volume report[192] about the voyage of the ship "Fram" in 1893–1896 that contains a wealth of new knowledge about the northern coast of western Siberia and the Arctic Ocean.

5.2.4. Conclusion

To use humans for purely scientific exploration of outer space is prohibitive, in terms of excessive cost as well as because of the high risk involved. For exploration in its classical meaning, i.e., travelling to unknown regions, human involvement is essential and the high cost and risk become bearable. Hence, science is not and cannot be a driver for human exploration of outer space. However, new scientific knowledge is, and will be, a natural by-product of humans going to the Moon, Mars, and beyond.

---

[185] Cosmic Vision spec. issue of Experimental Astronomy (2008). Please find the complete information for this footnote.

[186] Ball, Philip. "Space experiments should be done on the cheap". Nature News 24 September 2007: doi:10.1038/news070924-13. Specify the information on pages.

[187] Haerendel, Gerhard. "Exploration needs cooperation". Space Research Today 169 (2007): 32–34.

[188] Bothmer, Volker, and Daglis, Ioannis A. Space Weather: Physics and Effects. Berlin: Springer, 2007.

[189] Baumjohann, Wolfgang, and Rumi Nakamura. "Magnetospheric contributions to the terrestrial magnetic field". Treatise on Geophysics. Schubert, Gerald ed. vol. 5. Amsterdam: Elsevier, 2007. pp. 77–92.

[190] Lockwood, Mike. "Fly me to the moon?" Nature Physics 3 (2007): 669–671.

[191] Driesman, Andrew, Hynes, Shane and Cancro, George. "The STEREO Observatory". Space Science Review (2008).

[192] Nansen, Fridtjof. Farthest North. 2 vols. Westminster, 1897.

5.3 Humans leaving the Earth – a philosopher's view

Jacques Arnould

Fig. 1. *Pieter Bruegel, Landscape with the Fall of Icarus (c. 1558).*

My name is ποιμένας (the shepherd). I was born in Herakleion and I have never left Crete since. I'm a shepherd, like my father was before me. When all this happened, I was grazing my flock close to the shore. There's not much grass, and it tends to be salty, but at ploughing time there's not much good grazing to be found anywhere else; even the sheep seem to know that and don't complain. I didn't see the young chap drown. I just heard someone cry out something that sounded like "Father!" I didn't pay much attention; I was watching a flight of swallows, trying to tell whether they meant that rain was on the way or whether it would stay fine. But you should ask my friend ψαράς (the fisher) who was fishing down by the shore. Oh, I see you've already asked him.

Did I know the young man? Yes, of course – though perhaps I should say it was rather his father I knew, Mr. Daedalus. He's a proper gentleman, an intellectual come down here from Athens. What do I know about him? He works for King Minos. He came to me once because he wanted to buy a cowhide. He gave me a good price for it too. Icarus – as you probably know – was a son of his, born to a serving girl at the palace. It's a bad business, him dying so young like that. Since you're from the police, perhaps you can tell me: what happened to the poor lad, exactly? Did he throw himself off the boat that I saw anchored in the bay? Unrequited love, perhaps? He seemed no more than an adolescent. What did you say; he fell out of the sky! Come on, you're pulling my leg! Oh, all right, if you say so ... Well then, it must have been him I saw flying towards the sun just on midday. I thought it was an eagle at the time.

If you want my opinion, I wouldn't like to say he deserved what happened to him, but he was tempting providence. Whatever was he doing up there? The heavens are for gods and goddesses, not humans. If we were meant to fly like birds, we'd be born with wings. And what have we got instead? Feet for walking on the ground and hands to hold a stick, like mine. Don't you think we've got enough ways of getting around? I'm constantly on the go with my flock. We have to cover a lot of ground to always be on a good pasture at the right season. In the trade we call that transhumance: here today, gone tomorrow. Our hero, for us pastoral shepherds, is Ulysses: on the one hand, a ship to cross the seas and have exciting adventures; on the other, a wife waiting at home. Nothing like the crazy scheme of this poor Icarus chap; not surprising that he came to grief.

Oh, I see, you think he was trying to escape from the island! Well in that case, it must have been his father's decision – everyone at the palace knows he's not on very good terms with his employer; I heard the King even had father and son shut up in the labyrinth to stop them leaving. I can understand why Daedalus had wanted to get away: ever since the Queen went off the rails and gave birth to that monster, life has become a lot more difficult for us ordinary people. The worst thing that can happen is to have a cuckolded king! But is that any reason to run out on us? I think Daedalus has a lot to answer for, trying to get away and let us all down, rather than looking for a solution to help us all out of trouble. We give him a big welcome, make sure he lacks nothing and then one fine morning he ups and disappears. I sometimes think I'd like to get away from here too, but I tell you, I wouldn't just abandon everybody – I'd try and take my family and friends with me.

I must say though, when you think about it, it must be wonderful to be able to fly and see the world from above. A few years ago I decided to climb Mount Ida with a friend and colleague of mine. *We came across an old shepherd in one of the mountain dales, who tried, at great length, to dissuade us from the ascent, saying that some 50 years before he had, in the same ardour of youth, reached the summit, but had got nothing for his pains except fatigue and regret, as well as clothes and body torn by the rocks and briars. No one, so far as he or his companions knew, had ever tried the ascent before or after him. But since youth is suspicious of warnings, his counsels increased rather than diminished our desire to proceed. So the old man, finding that his efforts were in vain, went a little way with us, and pointed out a rough path among the rocks, uttering many admonitions, which he continued to send after us even after we had left him behind.* Well, it was hard going, I can tell you, and as we struggled up I often thought of the words of some poet, Ovid I believe, that our schoolteacher (oh yes, I went to school) made us learn by heart: "Wishing is not enough; to possess a thing fully, you must desire it ardently." Ah, we wanted it already, that wretched summit. It took us hours and hours, but we finally made it. *At first, owing to the unaccustomed quality of the air and the effect of the great sweep of view spread out before me, I stood like one dazed. I beheld the clouds under our feet, and what I had read of Athos and Olympus seemed less incredible as I saw with my own eyes the same things from a mountain of less fame. I turned my gaze towards* the part of the island *whither my heart most inclined.* I could see Herakleion, my parents, my family and my flocks away in the distance. *I sighed, I must confess . . . and an inexpressible longing came over me to see* them *again.*[193]

I can see you're surprised, Mr. Policemen. You didn't expect a shepherd from Herakleion to harbour feelings of nostalgia in his bosom, I suppose? Although our jobs have something in common; I reckon keeping watch over sheep is a lot easier than keeping watch over people and leaves more time to learn and especially to think. And don't forget, we've always got one eye on the heavens, watching the Evening Star. It's the first to come out at dusk and the last to die away at dawn. It shines just at the time we're either driving the sheep to the fields or bringing them home to the fold. That's why some call it the Shepherd's Lamp.

What's that, you say it's a planet? Venus? Oh, I didn't know. You mean she doesn't actually shine on her own but because of the sun? That seems a shame, but then again it's rather touching that our gateway, our guide to the heavens, should be a woman. Nature is feminine too, and I love nature. Besides, getting back to Icarus, I wonder whether those two (because you did say Daedalus had managed to fly as well, didn't you?) would have survived very long, up there in the sky. I mean all birds, even the ones that cross the seas, have to come down sometime. I've said it before, but up there is no place for us humans and animals. They would have had to take up a house to sleep in and a garden to grow things to eat: we're not like the gods, living on nectar and ambrosia! If ever one day – but I can't see how we would ever do it – we manage to navigate through the clouds and land on the moon, I bet the first thing we do will be to sow wheat and plant vines!

Really? That's exactly what my friend ζευγολάτης (the ploughman) told you? It's true, when the tragedy occurred he wasn't far off – he'd just finished ploughing his field. He didn't see anything either? I'm not surprised: when you're behind a horse and plough, you've no time to stare at the sky. What's more, I might as well

Ever since, expectations towards commercial benefits have been made also with respect to *manned* space activities in a somewhat strange consent of proponents and opponents of spaceflight. Consequently, the proponents of manned space activities developed industrial scenarios in the late 1980s, e.g., promising large-scale production of valuable materials under unique microgravity conditions of space in due time. However, these ambitions were based on rather optimistic expectations and have ended in disappointments in some cases. As a consequence, space "business" enthusiasm has turned into profound pessimism among the public regarding the significance of spaceflight, accompanied by more restrictive space policies. Manned spaceflight has not yet fully recovered from this setback.

The reason for this lasting disappointment lies in implicit, but unchanged utilitarian views of many decision-makers and actors, which suffer from *economy-driven myopia* in the assessment of future spaceflight options while ignoring the other cultural benefits of spaceflight apart from the monetary valuation. It is therefore not surprising that from this limited perspective even those arguments in favour of manned spaceflight, e.g., that stress the interactive role of humans in "intelligent" space utilization and exploration, are easily disputable from the same biased perspective. Thus, an argument put forth is that humans in space affect their specific space environments while working in it. This may be the case when astronauts threaten to interfere with their sensitive microgravity experiments or with pristine planetary surfaces, thus being blamed of risking the return of investments. In fact, these failures cannot be ruled out for robotic missions either, but the "moderate" costs here seem to be acceptable for the sceptics who prefer unmanned spaceflight. However, *any* option to act would hardly be attractive if poor results were to be expected, even assuming "only" moderate expenditures. This holds true unless there are non-utilitarian justifications of spaceflight. Any valuation of manned spaceflight as being generally "too expensive" would be arbitrary, because this statement lacks the necessary reference to the aspect of human aims that helps one to assess whether certain actions should be taken or not. The question therefore would be if culture matters beyond monetary terms.

5.4.3. The trans-utilitarian perspective

Spaceflight just like other grand challenges and long-term commitments on Earth cannot be assessed in monetary terms alone. The erection of the Eiffel tower or the operation of theatres might be indeed be considered irrational – but only with respect to the financial aspect. However, societies apply – and have applied in the past as well – non-utilitarian criteria for good reasons: to meet

Cultural Aims	Perspectives
Enhancement of the human cultural sphere towards a cosmic culture	World view; technology; human exploration & presence
Contribution to a polycentric world and a multilateral political culture on Earth	International cooperation & national share in new options; plurality of space-faring nations & balance of power
Returns for science	Knowledge; cognition (also from unmanned activities)
Benefit: Added economic value	Not yet foreseeable

Fig. 2. *The cultural dimension of human spaceflight (source: Stephan Lingner).*

specific needs or preferences irrespective of their substitution potential for money or other exchangeable goods. Actually, the conversion potential of the latter cannot be an aim in itself – it merely increases the flexibility of the actors and is therefore more relevant as a means rather than goals. Even if welfare or earning money were legitimate (secondary) goals, they should not exclude other aims as legitimate or rational. In fact, economic rationality is part of a broader cultural practise, which encompasses political, juridical and scientific rationality, among other things. The outcomes of culture are societal contracts, human cognition and development, which are central to humanity but hard to estimate in monetary terms. As regards the valuation of spaceflight, one might thus state that it has to be seen as a broader *cultural option* of humankind. Corresponding trans-utilitarian objectives are manifold and might be classified along several sub-categories[196] (see Figure 2).

Basically, space travel enables the *expansion of the human cultural sphere* by space exploration and presence in space as well as by corresponding technological progress[197] and finally, by the enhancement of human cognition.[198] From this perspective, spaceflight may be viewed in the tradition and as an extension of the fascinating major discoveries of *terra incognita* and deep-see regions in the past. Thus, the famous Alexander von Humboldt travelled through South America explicitly for the purposes of research and public enlightenment.[199] Ultimately, space travel will widen the scope of human options and choices – in the sense of advancing humanity – by establishing a cosmic culture. The latter – if desirable – would be even exclusive with respect to spaceflight as means (Figure 3).

Another line of argument is based on the maxim of a peaceful civilization on Earth: spaceflight could contribute to the *multi-lateral* organisation of global society by *international cooperation* and participation in its major projects. The

177

Fig. 3. *Manned exploration of the lunar surface, Apollo 17 (source: NASA).*

plurality of space-exploring nations might thus balance the distribution of power towards a polycentric political culture on Earth. This aim might be indeed achieved by other international enterprises too, but human spaceflight is a promising option for peaceful cooperation, here.

Although human spaceflight might also give new insights and knowledge in planetary science, astronomy and cosmology, it has to compete with corresponding unmanned activities, here. Finally, the creation of economic value by human spaceflight is yet not clearly foreseeable, although certain conceptions might promise benefits, e.g., production and utilization of power from space.

5.4.4. Arguments at national level

The above-mentioned cultural rationales for human spaceflight generally also apply to the national level. A specific cooperation motive for space-faring nations lies clearly in the expectation of national participation in the new options. Another highly critical motive for human spaceflight gave way to the "space race" during the Cold War. It aimed at the predominance of one society over another. This imperialistic motive cannot be generalized in a peaceful environment, because its notion violates the basic "neminem laede" principle (not-harming-others principle). Therefore, it is not acceptable for the community of states.

The meaning of the *leadership motive* seems to be somewhat similar if it were directed towards cementing the asymmetry of power for overwhelming others.

Nevertheless, leadership as an ambivalent notion and should not be blamed as biased and adverse in every case: instead, those leadership motives that are based solely on initiative, ability and even national prestige should be weighted as legitimate. In this respect and for practical reasons, leadership might be even necessary, because some countries have to take the lead in complex missions as long as not all interested countries are able to do so. All countries would be better off as a result, which is in line with the "difference principle" in ethics. Therefore, leadership could be justified by this notion, if we accept some differences within the context of this principle.

Another argument for already space-faring countries is "keep existing options open". Giving up any already acquired spaceflight know-how would need explaining. The reasoning would have to come from the opponents – not the proponents in this case. This would apply similarly to the closure – not the operation – of already existing theatres, operas and other cultural institutions.

5.4.5. Remaining ethical questions

In addition to this differentiated view, there are some remaining problems of justification of human spaceflight, which concern both utilitarian and trans-utilitarian perspectives on national and international levels:[200] The sceptical thesis that humans should be modest[201] instead of aiming for the stars is a sufficiency claim, which strictly speaking cannot be generalized nor it limitations defined.[202] Moreover, it would question all technological progress as well, and implicitly, the cultural development of humanity as a whole, which is hardly acceptable. This thesis is somewhat similar to the (often religiously motivated) ideal that human's destiny is restricted by its "natural" *terrestrial* borders, because either for the aforementioned reason of modesty or for reasons of the "sacrosanct" nature of space. This argument may be classified as a *naturalistic fallacy*, because the perception of systems of borders alone – like that of Earth's surface – cannot prescribe tolerable limits to human action. One should not forget that the European continent did not pose any "normative power" against Columbus or Magellan while preparing for new overseas discoveries or against early humans migrating "out of Africa".

A third sceptical argument stresses the hostility of the space environment for human life. Therefore, humans should refrain from spaceflight. This argument is not convincing with respect to the availability and potential of appropriate life-supporting systems. Humans have already proven their ability for technical adaptation to other harsh environments by settlements in Siberia, manned expeditions to the poles, to deserts and deep-sea regions as well as modern air-travel. In summary, there is no generally valid argument for human self-restraint that could be put forth against an ethos of transcendence towards space.

5.4.6. Conclusion

This short analysis argues that manned spaceflight as a principally legitimate and reasonable enterprise. However, this does not mean that it *must* be carried out as a priority nor that there is no need for case-sensitive, in-depth analyses of the mission-specific ethical questions.[203] In fact, public and private actors have to decide on the *extent* of their participation in international space exploration as *one* option among others, weighing all utilitarian *and* trans-utilitarian criteria, quite similar to the reasoned choices for or against other cultural enterprises on Earth.

Any decisions should respect the potential of human spaceflight for the enrichment of human opportunities and choices by overcoming natural obstacles of human cultural development. Human spaceflight might not necessarily have concrete benefits in the short-term, but will open the way to new ones. Nonetheless, decision makers should make their criteria and reasoning transparent and explicit to the public. This would be prudent – at least over the long run – for the *stable acceptance* by the public and would constitute a clear mandate for the potential long-term objective of international space exploration.[204]

Finally, the perceived dichotomy of utilitarian and trans-utilitarian valuation should not mislead opponents or competing actors of spaceflight to play-off one approach against the other for two reasons: Both approaches ultimately contribute to cultural development, which is driven by the two different albeit complementary aspects. Moreover, the two sides of the coin represent ends of an unbroken justification for spaceflight rather than a polarity of possible reasons, thus leaving no room for inherent weaknesses in the overall concept.[205]

---

[195] Examples may be large ion-colliders and electron–synchrotons for nuclear physical research.

[196] Gethmann, Carl Friedrich. "Manned Space Travel as a Cultural Mission". Poiesis & Praxis. International Journal of Ethics of Science and Technology Assessment 4 (2006): 239–252.

[197] This may also include "spin-offs" as desirable surprises.

[198] Another side-effect resulted from the well-known "overview effect" which became a driver for world-wide environmentalism for more the 30 years. See also Seboldt, Wolfgang et al. "A review of the long-term options for space exploration and utilisation". ESA Bulletin 101 (2000): 31–39.

[199] Knobloch, Eberhard. "Erkundung und Erforschung – Alexander von Humboldts Amerikareise". Poiesis & Praxis: International Journal of Ethics of Science and Technology Assessment 4 (2006): 267–288.

[200] Gethmann, Carl F. "Man in space: The Ethics of Space Policy". P. Pompidou, Alain ed. Paris: UNESCO. pp. 55–56.

[201] Jonas, Hans. Das Prinzip Verantwortung. Frankfurt a. M: Suhrkamp, 1979.

[202] Another quite similar but still invalid claim states that societal problems on Earth generally do not permit expenditure for ambitious spaceflight projects.

[203] Williamson, Mark. "Space ethics and protection of the space environment". Space Policy 19 (2003): 47–52.

[204] Wilholt, Torsten. "Scientific autonomy and planned research: the case of space science". Poiesis & Praxis. International Journal of Ethics of Science and Technology Assessment 4 (2006): 253–266.

[205] Schrogl, Kai-Uwe, and Nicola Rohner. "Für einen neuen Ansatz zur Begründung der Raumfahrt". Die Zukunft der Raumfahrt. Ihr Nutzen und ihr Wert. Gethmann, Carl Friedrich, Rohner, Nicola, Schrogl, Kai-Uwe eds. Bad Neuenahr-Ahrweiler: Europäische Akademie, 2007.

5.5   The need of a legal framework for space exploration

Ulrike M. Bohlmann[206]

5.5.1. Introduction

A growing number of exciting missions are being planned and undertaken in the exploration of our solar system.

What is the legal framework for these missions? Why is the framework as it is? Is it sufficient as it is? What developments are to be expected in the near future? Such are the questions this article attempts to answer. After giving a survey of the applicable legal regime with a special emphasis on the issues of non-appropriation, planetary protection, the use of nuclear power sources, and international coopera-

Fig. 4. *Artist's illustration of the proposed roadmap for ESA's Aurora exploration programme that could lead to a human flight to Mars (source: © ESA – P. Carril).*

tion, this framework will be assessed with a view to establishing some trends in the evolution of the law, which will lead us to some tentative conclusions on what future developments might be expected.

Space exploration initiatives continue to be a high-risk venture from several perspectives; as with all space activities, they require major investments and are still regarded as hazardous. The danger of accidents is relatively high and the potential damages are difficult to assess. Therefore, a viable legal framework is needed to ensure a balance of interests between the different groups of actors involved. Conflicting interests need to be harmonized into a compromise that achieves an equitable sharing of liabilities and benefits among all parties involved.

5.5.2. The term "exploration" in the *Corpus Iuris Spatialis*

The term "exploration" occupies a prominent place in the international space law codifications. The full title of the so-called "Magna Charta of Space Law" reads

Fig. 5. *An artist concept of ESA's ExoMars rover on Mars under study on behalf of the Aurora programme (source: © ESA-Medialab).*

"Treaty on Principles Governing the Activities of States in the Exploration and Use of Outer Space, including the Moon and Other Celestial Bodies".[207] The second paragraph of the preamble recognizes the common interest of all mankind in the progress of the exploration and use of outer space for peaceful purposes; Articles I and 2 stipulates the general freedom for all states to explore and use outer space, including the moon and other celestial bodies. In general, the term "exploration" signifies investigation, search, study, or travel for discovery parallel to a geographic expedition. In a narrower sense, the term is understood to mean investigation of the universe beyond the Earth's atmosphere by means of manned and unmanned spacecraft.

The use of the term "exploration and use" throughout the Outer Space Treaty also embraces the "use" that can be made of outer space in the sense of using it for other purposes than mere study and research, e.g., space applications. The fundamental legal basis of all space exploration activities can be found in Articles I, 2 and 3 of the Outer Space Treaty:

"Outer space, including the Moon and Other Celestial Bodies, shall be free for exploration and use by all States without discrimination of any kind, on a basis of equality and in accordance with international law, and there shall be free access to all areas of celestial bodies.

There shall be freedom of scientific investigation in outer space, including the Moon and Other Celestial Bodies, and States shall facilitate and encourage international cooperation in such investigation."

Addressees of the rights and obligations of the Outer Space Treaty are first and foremost the States Parties to the Treaty. Apart from the States as the sovereign political entities that concluded the Treaty other subjects, with or without legal personality under public international law, emerge from the text. The Treaty is *"inspired by the great prospects opening up before mankind"*. The *"common interest of all mankind in the progress of the exploration and use of outer space for peaceful purposes"* is recognized and the exploration and use of outer space shall be the province of all mankind the same as astronauts shall be regarded as the envoys of mankind. Furthermore, the exploration and use of outer space should be carried on for the benefit of all peoples irrespective of the degree of their economic or scientific development.[208] These references to mankind which is to be understood to mean the human race in its entirety and to all peoples of the Earth are part of the maxims that guided States Parties in the drafting of the Treaties. From a purely legal point of view, however, public international law, including the Outer Space Treaty, only addresses States in their relations and interactions. That is also the reason why any private initiative is always linked to a State Party in the system of public international space law.[209] The freedom to explore and use outer space is wide-ranging, but not unlimited. The most important limits to this freedom that are relevant to the subject of this study are the principle of non-appropriation, the protection of the environment and requirements for international cooperation.

5.5.3. The non-appropriation principle

The primary limit of the general freedom to explore outer space is the principle of non-appropriation as enshrined in Article II of the 1967 Outer Space Treaty.[210] Non-appropriation is the necessary corollary to the general freedom to explore outer space, since that freedom is only conceivable if territorial sovereignty is banned. Consequently, any means of appropriation is prohibited by the Outer Space Treaty. The principle of non-appropriation confirms that all nations are vested with equal rights and enjoy an equal access to space resources, regardless of their current degree of technological development.[211] Interesting is also the solution found in the Moon Agreement.[212] Article 11.2 reproduces exactly the same wording with respect to the Moon as can be found in Article II of the Outer Space Treaty. This general non-appropriation principle experiences a concretisation in Article 11.3 that states that

> *"Neither the surface nor the subsurface of the Moon, nor any part thereof or natural resources in place, shall become property of any State, international intergovernmental or non-governmental organization, national organization or non-governmental entity or of any natural person. [...]"*

The authors of the Moon Agreement were very explicit about the all-embracing applicability of the non-appropriation principle. Moreover, they were cautious to distinctly draw the line between the question of appropriation and property on the one hand and the right to collect and remove samples for scientific purposes on the other hand. Article 6.2 provides that

> *"In carrying out scientific investigations and in furtherance of the provisions of this Agreement, the State's parties shall have the right to collect on and remove from the Moon samples of its mineral and other substances. Such samples shall remain at the disposal of those States Parties which caused them to be collected and may be used by them for scientific purposes. [...]"*

The last sentence of Article 6.2 goes one step further in stipulating that

> *"States Parties may in the course of scientific investigations also use mineral and other substances of the Moon in quantities appropriate for the support of their missions."*

185

5.5.4. Protection of the environment

The implementation of space exploration initiatives also has an impact on the environment. Regarding the protection of the environment,[213] regulations exist on quite different levels and with different degrees of intensity. The provisions contained in the classical *corpus iuris spatialis*, above all Articles IX and XI of the Outer Space Treaty, remain ambiguous and leave any "appropriate measures" to be adopted by the States Parties to avoid the harmful contamination of celestial bodies and also adverse changes in the environment of the Earth at the sole discretion of the respective Party to the Treaty. The Moon Agreement, especially its Articles 7 and 4 elaborates a bit more on the principles regarding the protection of the extra-terrestrial environment by addressing the issues of forward and back contamination as well as the principle of intergenerational equity, and reflects thereby the fact that at the time it was drafted, in the late 1970s, environmental considerations had become a global concern.

Based on the policy statement, that

> *"Although the existence of life elsewhere in the solar system may be unlikely, the conduct of scientific investigations of possible extraterrestrial life forms, precursors, and remnants must not be jeopardized. In addition, the Earth must be protected from the potential hazard posed by extraterrestrial matter carried by a spacecraft returning from another planet."*

COSPAR, the Committee on Space Research,[214] has elaborated a detailed Planetary Protection Policy. The COSPAR Planetary Protection Policy is intended for the reference of space-faring nations, to guide compliance with the obligations of States Parties to the Outer Space Treaty. It is based on the policy of probabilistic avoidance of contamination. Five different categories are established for target body/mission type combinations and respective suggested ranges of requirements, based on the degree of interest they represent for the understanding of the process of chemical evolution or the origin of life. The COSPAR Planetary Protection Policy is a very consistent and highly developed system of recommendations by an independent and international body of scientists with the highest reputation in the field. However, the price to pay for the specificity and wealth of detail of these guidelines is their lack of legal force; they are not legally binding, since COSPAR is a non-governmental organisation without institutionalized authority. Still, it is the continuous policy of many actors in the space field to comply with COSPAR's recommendations and to model their national or internal planetary protection policies according to COSPAR standards. One of the most prominent examples is the NASA Planetary Protection Policy. The very detailed and elaborated NASA Planetary Protection Guidelines are intended to apply not

Fig. 6. *From Earth to Mars via the Moon (source: © ESA – Estudio IADE).*

only to NASA missions but also to the flight of NASA instruments or experiments on non-NASA spacecrafts. However, the quality of these policies and guidelines remains that of a national internal document that is not binding internationally.[215] The ESA Planetary Protection Policy has just recently been adopted by the ESA Council on 10/11 October 2007. It is also compliant with the COSPAR planetary protection policy and the corresponding implementation guidelines and stipulates that spaceflight missions carried out with any degree of ESA involvement shall comply with this policy and its associated requirements.

The elements of general international environmental law, which need to be mentioned in the context of this study, are Principle 21 of the Stockholm Declaration,[216] Principle 15 of the Rio Declaration[217] and the Convention on Biological Diversity.[218]

According to Principle 21 of the Stockholm Declaration:

> "*States have, in accordance with the Charter of the United Nations and the principles of international law, [. . .] the responsibility to ensure that activities within their jurisdiction or control do not cause damage to the environment of other States or of areas beyond the limits of national jurisdiction.*"

Thus, the environment of outer space as one of the areas beyond the limits of national jurisdiction is protected by this principle. The United Nations General Assembly Resolution 2996 (XXVII) 1972 asserts that Principle 21 [and 22] of the Stockholm Declaration 'lay down the basic rules governing the matter'.[219] Principle 2 of the Rio Declaration and Article 3 of the Convention on Biological Diversity repeat the Principle, so that – although the Stockholm Declaration has no legally binding character – at least Principle 21 may be regarded as reflecting customary international law. Principle 15 of the Rio Declaration elaborates further on the so-called precautionary approach[220] but is significantly weakened by the reference to the respective States' capabilities. The essence of the precautionary approach as contained in Principle 15 of the Rio Declaration is best described by *P. Birnie and A. Boyle*:[221] in performing their obligations of environmental protection, states cannot rely on scientific uncertainty to justify a lack of action when there is enough evidence to establish the possibility of a risk of serious harm, even if there is no proof of harm.

Article 8 (h) of the Convention on Biological Diversity provides that

> "*Each Contracting Party shall, as far as possible and as appropriate prevent the introduction of, control or eradicate those alien species which threaten ecosystems, habitats or species*".

According to Article 4 of the Convention, its provisions apply, in relation to each Contracting Party, regardless of where the effects of activities occur, when carried out under the jurisdiction or control of a Contracting Party, within the area of its national jurisdiction or beyond the limits of national jurisdiction. Thus, it is also applicable to the outer space activities of Contracting Parties.[222] Accordingly, the issues of forward and back contamination are dealt with by the Convention; the potentially harmful introduction of species that are foreign to a given environment, be it terrestrial or extra-terrestrial, is to be prevented, or at least controlled or eradicated. This obligation is, however, limited to "as far as possible and as appropriate".

The short overview of public international law that is relevant to planetary protection issues shows the influence a new kind of thinking in ethical and environmental terms has had on the development of the law. The early texts, like the Outer Space Treaty and the Liability Convention, base themselves on anthropocentric and geocentric values. The sole reason for the protection of the extraterrestrial environment was the preservation of scientific opportunities. The COSPAR Planetary Protection Policy follows along the same lines. The Moon Agreement elaborates further on the balance of the extra-terrestrial environment, and mentions the principle of intergenerational equity for the first time in space law. Intergenerational equity is one of the main subjects of modern international environmental law, based on the Rio Declaration and geared towards the avoidance of irreversible harm. Nevertheless, the Moon Agreement still lays special emphasis on the scientific interest, for which international preserves might be established. The idea to conserve species and their habitat for their intrinsic value and not just as resources exploitable by man is still relatively young; in the Convention on Biological Diversity a shift is starting from an anthropocentric to an eco-centric or bio-centric ethical approach that attributes value to all life as part of an ecosystem or to all life as such. The next stage of development, to cosmo-centric ethics recognizing the intrinsic value of the extraterrestrial environment and of its existing balance has not yet occurred.

5.5.5. The use of nuclear power sources

For a number of technical reasons, nuclear power sources constitute the only viable option for power supply on most exploration missions, a fact that is already recognized in the preamble of the "Principles Relevant to the Use of Nuclear Power Sources in Outer Space".[223] According to Principle 3, individuals, popula-tions, and the biosphere are to be protected against radiological hazards and the contamination of outer space is to be avoided. Sections 2 and 3 of Principle 3 establish specific rules for the use of nuclear reactors, on the one hand, and radioisotope generators, on the other. Principle 3 Section 2 thus allows – with some further restrictions as to the fuel to be used, the design, and the construction of the reactor – the operation of nuclear reactors on interplanetary missions, in sufficiently high orbits and in low-Earth orbits, if the reactor is stored in a sufficiently high orbit after the operational part of the mission. Principle 3 Section 3 allows the use of radioisotope generators – under certain technical and design conditions – for interplanetary missions and other missions leaving the gravity field of the Earth. They may also be used in Earth orbit if, after the conclusion of the operational part of their mission, they are stored in a high orbit. In any case, ultimate disposal is necessary. Principle 4 stipulates that a launching Sate has to ensure that a thorough and comprehensive safety assessment is conducted. The results of this assessment shall be made publicly available prior to each launch. Furthermore, the Principles contain provisions as to the notifi-cation in case of re-entry of satellites with nuclear power sources on board, Principle 5. Principle 9 concretises Article VII of the Outer Space Treaty and the Liability Convention: it affirms that international liability fully applies to cases where a space object carries a nuclear power source and provides that the compensation to be paid shall be determined in accordance with international law and the principles of justice and equity, in order to provide reparation in respect of damages to restore the person, natural or juridical, State or international organization on whose behalf a claim is presented to the condition which would have existed if the damage had not occurred. Compensation includes reimbursement of the duly substantiated expenses for search, recovery and clean-up operations. The Principles Relevant to the Use of Nuclear Power Sources in Outer Space are currently under review. The Scientific and Technical Subcommittee of the UN Committee on the Peaceful Uses of Outer Space is working on the development of goals and recommendations for the safety of NPS applications in outer space. Focused studies and a partnership with the International Atomic Energy Agency are intended to allow the examination of a range of issues with a view to recommending a procedure for establishing technical safety standards.

5.5.6. International cooperation

Article I, 1 of the Outer Space Treaty, declares the exploration and use of outer space to be the 'province of all mankind' that shall be carried out for the benefit and in the interest of all countries. This formulation has been seen as inducing a change in the orientation of public international law from a law of mere co-existence to a law aiming at cooperation. It strikes a note of solidarity, which embraces a prohibition of monopolisation of products resulting from space activities for national purposes; the striving for the establishment of equal possibilities to use outer space, and the postulate to implement space activities by means of co-operation wherever possible.[224] We also need to mention the *Declaration on International Cooperation in the Exploration and Use of Outer Space for the Benefit and in the Interest of All States, taking into Particular Account the Needs of Developing Countries*,[225] which underlines in its second operational paragraph the freedom of States to cooperate and to determine all aspects of such cooperation as well as the requirement to organise cooperation in an equitable manner and emphasises in its third operational paragraph that international cooperation shall take place on a basis that is acceptable for all Parties concerned.

Due to the enormous ambitions connected with space exploration initiatives, international cooperation is a key issue in their implementation. Therefore, 14 space agencies have jointly developed The Global Exploration Strategy: The Framework for Coordination[226] as a vision for globally coordinated space exploration. It elaborates an action plan to share the strategies and efforts of individual nations so that all can achieve their exploration goals more effectively and safely. The strategy allows for optional participation based on the level of interest at each agency. It introduces an international coordination mechanism as a voluntary non-binding forum through which agreement can be reached on interoperability

Fig. 7. *An artist concept of robotic activities at a manned lunar station based on technologies to be developed under ESA's Aurora space exploration programme (source: © ESA-Medialab).*

standards for practical features such as communications, control, life support and docking systems. It aims at developing mechanisms for provision of payload opportunities and at providing a forum to discuss issues such as technology transfer and property rights.

A last point that needs to be mentioned is the status of astronauts as envoys of mankind. According to Article V of the Outer Space Treaty, States Parties to the Treaty shall regard astronauts as envoys of mankind in outer space and shall render to them all possible assistance in the event of accident, distress, or emergency landing. In carrying on activities in outer space and on celestial bodies, the astronauts of one State Party shall render all possible assistance to the astronauts of other States Parties. States Parties to the Treaty shall immediately inform the other States Parties to the Treaty or the Secretary-General of the United Nations of any phenomena they discover, which could constitute a danger to the life or health of astronauts. These principles have been further elaborated in the Agreement on the Rescue of Astronauts, the Return of Astronauts and the Return of Objects Launched into Outer Space. The so-called 1968 Rescue Agreement develops and gives further concrete expressions to these duties. Article 10 of the Moon Agreement goes still another step further in proclaiming that: "States Parties shall adopt all practicable measures to safeguard the life and health of persons on the Moon" and thereby extends the treatment owed to astronauts to any person on the Moon. Furthermore, States Parties shall offer shelter in their stations to persons in distress. All these regulations are founded upon a commitment to international cooperation in the peaceful exploration and use of outer space, and upon recognition of the need for international cooperation in responding to accidents, emergencies or other forms of distress. They elaborate on a humanitarian notion.

5.5.7. Some trends in the evolution of the law

A number of hypotheses can be deduced from this tableau of norms and rules, binding and not-so-binding. We have distinguished three phases in traditional space law-making in the United Nations setting:[227] The first phase saw the successful elaboration of the fundamental international space law conventions. Work at the UN was driven at the time by the space race, that is, by national interests concerning power maximisation and survival. Binding commitments were taken, even though the language of the Treaties remains deliberately vague. The second phase of space law-making was marked by the adoption of special legal regimes in form of UN General Assembly Resolutions, as the ones on Direct Broadcasting, or the Remote Sensing Principles. Now, we have entered into a third phase that – due to a growing number and diversity of space-faring nations and entities – stumbles into a new flexibility. Since the international community as such has not yet succeeded in developing legal instruments containing binding commitments, the scientific community is introducing its own "rules of the road", as for example the COSPAR Planetary Protection Principles. Developed in the consensual culture of the scientific community, where international cooperation is considered natural and welcome, these "Rules of the Road" leave aside the politicians' national prestige considerations. The law concerning space exploration,[228] and space activities in general, is therefore characterized by a combination of – on the one hand, binding – Treaties and Agreements with deliberately imprecise language, and, on the other hand, detailed and very specific non-binding standards and guidelines.

This legal status quo reflects a compromise between opposite and conflicting positions. The rules and regulations from the early age of space law-making attempt to strike a balance between the general freedom to explore and use outer space by the current space powers, and, the wish to guarantee those same freedoms to States not yet capable of exercising these legal freedoms as well as to ensure the benefits of such activities to future generations. Consideration is also given to the protection of the extraterrestrial and terrestrial environment, while putting a special emphasis on the scientific concerns. This balance is fragile because most of the rules establishing the legal framework are considered non-committing soft-law instruments. The possibilities for international control are limited. The scientific community gives an important impulse for the development and operation of space exploration programmes and also plays a central role in developing the new "rules of the road", because even though the willingness of States to commit themselves internationally seems to be decreasing, we are witnessing a tendency of States to voluntarily accept non-binding international standards and guidelines as a basis for their own national or internal policy and legislation. The respective ESA and NASA Planetary Protection Policies[229] based on the COSPAR model are cited as examples.

In sum, the evolution of the law governing space exploration towards an increased influence by the scientific community reflects also the shift of political motivations for these space exploration initiatives from the early hard power arguments to the quest for scientific knowledge perceived as a cultural imperative.[230]

---

[206] The views expressed in this article are purely personal and do not necessarily reflect the views of any entities with which the author may be affiliated.

[207] The Outer Space Treaty, 610 UNTS 205, was adopted by the UN General Assembly in its resolution 2222 (XXI). It was opened for signature on 27 January 1967 and entered into force on 10 October 1967. As of 1 January 2008, it received 98 ratifications and 27 signatures. Thus, it can be regarded as having universal legal value.

[208] See also chapter 6 on international cooperation.

[209] Article VI of the Outer Space Treaty stipulates that States Parties shall bear international responsibility for national activities in outer space, regardless of whether such activities are carried on by governmental agencies or non-governmental entities, i.e. private parties. States Parties are responsible for ensuring that national activities are carried out in conformity with the provisions of the treaty and the activities of non-governmental entities in outer space shall require the authorization and continuing supervision by the appropriate State party to the Treaty.
See also on the "private trade" in "lunar deeds" footnote 210.

[210] It reads: "Outer space, including the Moon and Other Celestial Bodies, is not subject to national appropriation by claim of sovereignty, by means of use or occupation, or by any other means." This excludes logically also any claim by any private party to own the Moon, other celestial bodies or parts thereof since the prohibition of national appropriation also precludes the application of any national legislation on a territorial basis to validate a 'private claim'. Therefore, the trade in "deeds to lunar property" cannot convey any recognized right or legal title, see the recent statement by the Board of Directors of the International Institute of Space Law, IISL, on Claims to Property Rights regarding the Moon and Other Celestial Bodies, available at http://www.iafastro-iisl.com/additional%20pages/Statement_Moon.htm.

[211] For a more detailed analysis of the principle of non-appropriation see, Ulrike M. Bohlmann, Legal Aspects of Space Exploration Initiatives, in: Benkö and Schrogl, Essential Air and Space Law, Utrecht: Eleven International Publishing Inc., 2005. pp. 215–240, under 2.

[212] According to Article 1, paragraph 1 of the Moon Agreement "The provisions of this Agreement relating to the moon shall also apply to other celestial bodies within the solar system, other than the earth, except in so far as specific legal norms enter into force with respect to these celestial bodies". However, since the Moon Agreement has received only thirteen ratifications and four additional signatures (status as of 1 January 2007), its force and value are rather limited; no customary value can be attributed to its regulations and it is only binding upon its States Parties.

[213] See on this also Ulrike M. Bohlmann, Planetary Protection in Public International Law, presented at the 46th Colloquium on the Law of Outer Space, October 2003, Bremen, Germany, published by the American Institute of Aeronautics and Astronautics, in the Proceedings of that Colloquium. 18 pp.

[214] COSPAR, the Committee on Space Research, was established in October 1958 by the International Council of Scientific Unions, ICSU, to continue the co-operative programmes of rocket and satellite research undertaken during the International Geophysical Year (1957–1958). Its objectives are to promote on an international level scientific research in space, with emphasis on the exchange of results, information and opinions, and to provide a forum, open to all scientists, for the discussion of problems that may affect scientific space research.

[215] According to the NASA Planetary Protection Officer, J. Rummel, the policy is based on the desire to preserve extraterrestrial environments for the science opportunities; cited after L. Woodmansee, If Life exists on Mars, our robotic probes may have brought it there, http://www.spacedaily.com/news/life-01zg1.html; see also Rummel, J. Planetary exploration in the time of astrobiology: Protecting against biological contamination, in: Proceedings of the National Academy of Sciences, available at: http://www.pnas.org.

[216] Declaration of the United Nations Conference on the Human Environment, adopted in Stockholm on 16 June 1972, 11 ILM 1416.

[217] Declaration of the United Nations Conference on Environment and Development, adopted in Rio de Janeiro on 12 August 1992, 31 ILM 874.

[218] 31 International Legal Materials 818 (1992), opened to signature on 22 May 1992 and entered into force on 29 December 1993.

[219] 112 States voted in favour of this resolution, none opposed, the then Eastern Bloc States abstained on Res. 2996, but have supported subsequent treaties recognizing the normative character of Principle 21.

[220] The Principle reads: "*In order to protect the environment, the precautionary approach shall be widely applied by States according to their capabilities. Where there are threats of serious or irreversible damage, lack of full scientific certainty shall not be used as a reason for postponing cost-effective measures to prevent environmental degradation.*" See also: Paul B. Larsen Application of the precautionary principle to the Moon, in Journal of Air Law and Commerce, Spring 2006. p. 295.

[221] International Law and the Environment, 2nd ed. Oxford, 2002. 120 pp.

[222] With the exception of the United States of America, all space-faring nations are Party to the Convention.

[223] They were adopted unanimously by the UN General Assembly in its Resolution 47/68 of 14 December 1992 thereby obtaining universal acceptance, even though they do not create binding commitments under public international law. For a general overview of international law regarding nuclear energy, see: Mr. Elbaradei, Nwogugu, E. and Rames, J. International law and nuclear energy: overview of the legal framework, available at: http://ecolu-info.unige.ch/colloques/Chernobyl/pages/Opelz.html, where the authors also sketch the picture of a mix of legally binding rules and agreements on the one hand and advisory standards and regulations on the other hand. The authors describe the evolution of non-binding standards to binding commitments. See also, more specifically M. Benkö, Nuklearenergie im Weltraum, in: Böckstiegel, K.-H. ed. Handbuch des Weltraumrechts. Köln, Berlin, Bonn, München, 1991. pp. 457–475.

[224] Hobe, S. Die rechtlichen Rahmenbedingungen der wirtschaftlichen Nutzung des Weltraums, Berlin, 1992. 112 pp. It is interesting to note that here again, the Moon Agreement goes some steps further and shows a higher degree of concretization, see the last but one sentence of its Article 6.2 *"States Parties shall have regard to the desirability of making a portion of such samples available to other interested States Parties and the international scientific community for scientific investigation."* Article 5 of the Moon Agreement aims at securing a flow of mutual information and suitable coordination of simultaneously planned activities.

[225] Adopted by the UN General Assembly on 13 December 1996, UN Doc. A/RES/51/122, available at: http://www.oosa.unvienna.org/SpaceLaw/gares/index.html; M. Benkö/K.-U. Schrogl, The UN Committee on the Peaceful Uses of Outer Space: Adoption of a Declaration on "Space Benefits" and other Recent Developments, in Zeitschrift für Luft- und Weltraumrecht 1997, p. 228; See also on the foregoing developments: Hobe, S. Common Heritage of Mankind – an outdated Concept in International Space Law?, in: IISL, Proceedings of the 41st Colloquium on the Law of Outer Space. Melbourne, Australia, 28 Sept.–02 Oct. 1998. pp. 271–278.

[226] The text is available at: www.globalspaceexploration.org.

[227] See: Benkö, M. and Schrogl, K.-U. in their introduction to: Space law current problems and perspectives for future regulation, Utrecht: Eleven International Publishing Inc., 2005. p. 1.

[228] A more comprehensive analysis can be found by Bohlmann, Ulrike M. Legal Aspects of Space Exploration Initiatives, in: Benkö and Schrogl, Essential Air and Space Law, Utrecht: Eleven International Publishing Inc., 2005. pp. 215–240.

[229] See footnote 215.

[230] See also: Bohlmann, U. and Martinez, L. Fly me to the Moon, Legal and Political Considerations of Space Exploration Initiatives, presented at the 49th Colloquium on the Law of Outer Space, October 2006, Valencia, Spain, published by the American Institute of Aeronautics and Astronautics, in the Proceedings of that Colloquium, 117 pp. For an in-depth analysis of the shift in focus with regard to regime-change in the legal system for outer space, see: L. Martinez, Science in Service of Power: Space Exploration Initiatives as Catalyst for Regime Evolution, in Journal of Air and Space Law, Volume XXXII/6, November 2007. p. 431.

CHAPTER 6

THIRD ODYSSEY: HUMANS MIGRATING THE EARTH: HOW WILL IT AFFECT HUMAN THOUGHT?

6.1 Summary

Jean-Claude Worms

Let us project ourselves into the future and suppose that man has indeed developed the means to really leave Earth. Not just leaving it to achieve some flea hops to the Moon and come back. No, let us feature ourselves at a point in time when permanent bases have been built and are being used on the Moon, perhaps on a few nearby asteroids and even on Mars. These first bases should be peopled by astronauts, scientists and engineers, medical doctors and psychologists, doing science and technology work, maintenance chores, and essentially striving for survival. For the sake of argument let us accept that, in the mind of some people, it will be time for more: driven by curiosity and in order to extend their opportunities, humans may eventually search for settlements outside our planet. What is unimaginable today may become necessary in the future. Should technology enable humankind to do it, we can contemplate human settlements on the Moon and on Mars, trying to mould these arid and desert worlds into liveable places. Families will migrate off the Earth towards new homes; babies will be born, in space or on another planet of our solar system.

Another reason can come to mind: there are reasonable persons, highly respected scientists even, who truly believe that this might become a necessity: in Stephen Hawking's words (September 2007), "... *life on Earth is at the ever-increasing risk of being wiped out by a disaster such as sudden global warming, nuclear war, a genetically engineered virus or other dangers ... I think the human race has no future if it doesn't go into space ... I don't think the human race will survive the next 1000 years unless we spread into space*".

So whatever the reasons, for better or for worse, humankind leaving the Earth would truly mark the beginning of a new era. Furthermore, either through the discovery of life elsewhere in the solar system (extinct or extant), or through the reception of extraterrestrial radio signals, another new era will begin should humans realize they are not alone in the universe. Such discovery may likely cause the development of a new collective identity for humanity. What people believe in, and how such beliefs are structured, has a strong binding force on societies, on Earth and eventually beyond. Human belief systems, whether religious or secular, change in the context of new living environments, and in contact with other forms of life and societies. The humanities and the social sciences will gain in importance to help humankind to adapt to these new paradigms. Past encounters that took place on Earth show that human beings did eventually adapt to unforeseeable realities, although most of the time at very great costs.

It was thus the goal of this session to try and address a few of the issues that humankind should confront if these new eras really come into focus. Three speakers shared that challenging task.

Gerda Horneck first addressed the concept of habitability in the solar system by defining the notion of habitable zone, i.e., where liquid water has been present over extended periods. Mars lies just on the border of that zone, indicating that Martian habitability is highly controversial, especially since habitability conditions are very much different for micro-organisms and for humans. Horneck assumes a wet past for Mars, making it very similar to the early Earth from the standpoint of habitability. The present surface and atmosphere of Mars are, however, extremely hostile to life in general and certainly to human life. A putative Martian biosphere is therefore either extinct, or retracted into rare oases. For future missions to Mars, major critical items will be radiation hazards during the trip and on the surface of Mars, low atmospheric pressure, dust-related problems, effects of the lower gravity, and isolation. To enable such missions a broad research portfolio must therefore be implemented, as addressed in the HUMEX study roadmap. Horneck concluded on the vital importance of implementing stringent planetary protection guidelines, both to protect a putative Martian biosphere ("planetary parks"), and to protect Earth from possible backward contamination.

Paolo Musso then attempted to reflect on the potentially difficult confrontation of earthly religions with evidence of intelligent alien life. The exercise was actually limited to the catholic religion which would certainly make it necessary to extend in the future this reflexion beyond that much too restrictive border. P. Musso remains at the level of currently acceptable physics: he postulates that contact with such a civilization could only be achieved via electromagnetic waves, but apparently basing more his argument on the pure logic of the 57-year-old "Fermi paradox" revisited in 2002 (either there are no ETs, or interstellar travel is impossible), than on physics itself. This assumption of course ensures that the "encounter" will occur without trade, wars, and spread of disease commonly experienced in earthly encounters between civilizations: no Indians or buffalos of outer space are thus accounted for in this model. He then establishes a curious equivalence between rationality on the one hand, and morality and metaphysics on the other hand, to ascertain that ET, if it exists, is also a moral (or immoral) being and hence, is also able to understand our religious values. An interesting aspect of this formal reflexion is the author's view that, despite an official "wait and see" attitude by their church, many Christian people already believe in the existence of ETs so that their faith would not be shattered by a confirmation thereof. Actually Musso's conclusion rings like what would then become an assumed zealous appeal to spread the gospel in the Galaxy: after all, Christ was born in a remote suburb of a far-away city of the Eastern Roman Empire, just as Earth lies at the unfashionable edge of our galaxy . . .

Finally Debbora Battaglia comes in at an unusual angle into the issue of "ET diplomacy", on the armature of which she builds an interesting "anthropological model of visits". Embracing a (hers) – space – ethnographer's point of view, she proceeds to analyze how actual life in space is influenced by our Earth-bound imaginaries and political agenda, in part by looking in minute details at the diary of Soviet cosmonaut Valentin Lebedev in which he described his life on board the Soyuz 7 spacecraft. Battaglia follows Lebedev in his challenge to connect with the "alien" environment he finds himself in during 211 days, and as a matter of consequence for shaping an ethics of space diplomacy. From that analysis and that of the U.S. myth of indomitability transposed to outer space she draws the conclusion that ET diplomacy would favour an "alien-to-alien" type of exchange, since both parties in such an exchange would demonstrate vulnerability vis-à-vis the other, thus changing their ways sufficiently as to become slightly alien themselves.

6.2 Mars as a place to live?
Past, present and future

Gerda Horneck

6.2.1. Introduction

When tackling the question of what defines a place to live, one normally considers the chemical, physical or social conditions of an environmental envelope that allows growth and propagation for a given organism or group of organisms. While the environmental tolerances of microorganisms – the only habitants on Earth for the first 2 billion years of the history of life – cover a very wide range, the environmental border lines for humans are much narrower order to make a planet habitable – at least for microbial-type of life – a minimum of environmental requirements need to be considered; these are (i) carbon-based chemistry, (ii) adequate energy sources, as well as (iii) water in its liquid phase.

On the assumption that liquid water is essential for life, the common definition of a "habitable planet" has been one that can sustain substantial liquid water on its surface. Assuming a tolerable temperature range between about 0 and $+100°C$ at the surface of a planet, our solar system provides a habitable zone in an orbit between 0.7 and 2.0 AU (Astronomical Unit). In a more conservative estimate, the width of the habitable zone is restricted to the range of 0.95–1.37 AU (Figure 2). Venus, Earth and Mars are situated within this habitable zone or close by. With a mean distance to the sun of 1.52 AU, Mars is located at the outer border of the habitable zone in our solar system.

Environmental factor	Humans[1]	Microorganisms
Temperature	15–35°C	−20–113°C
Pressure	700–5000 hPa	<50–10,000 hPa
Radiation[2]	1–3 Gy	4000 Gy
pH	Near neutral	0–13
Oxygen	15–25%	0–100%
Carbon dioxide	<1%	0–100%

Fig. 1. *Physical and chemical environmental factors that define "a place to live" for humans or microorganisms (modified from*[231]*) [1]Environmental conditions for humans without any technical aids. [2]Sub-lethal dose.*

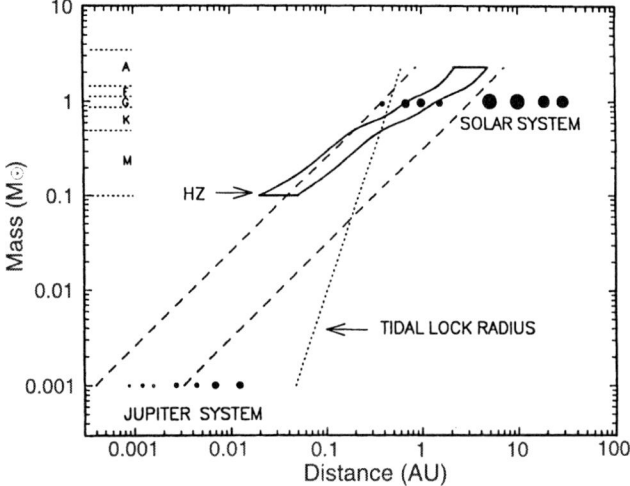

Fig. 2. *"Habitable zone as a function of the distance from the star and its mass" (source: Franck, Siegfried, and 5 co-authors. Habitable Zones in Extrasolar Planetary Systems. in: Astrobiology, the Quest for the Conditions of Life. Gerda Horneck, and Christa Baumstark-Khan, eds. Berlin Heidelberg New York: Springer, 2002. pp. 47–56).*

With the exception of the Earth, Mars is by far the most intensively studied of the planets of our solar system. In 1972, a spacecraft, Mariner 9, passed over the younger parts of Mars for the first time revealing a wide variety of geological

Environmental element	Mars	Earth
Cosmic ionizing radiation	100–200 mSv/a	1–2 mSv/a
Solar particle events	up to~0.1 Sv/h	not applicable
Solar UV radiation (spectrum)	$\lambda \geq 200$ nm	$\lambda \geq 290$ nm
Solar constant	589.2 W/m^2	1367.6 W/m^2
Length of day	24 h 37'22.7"	23 h 56'4.1"
Gravity	$0.377 \times g$	$1 \times g$
Atmosphere	95.3% CO_2	78.1% N_2
	2.7% N_2	20.9% O_2
	1.6% Ar	0.03% CO_2
	0.1% O_2	
Mean pressure	560 Pa	10^5 Pa
Average surface temperature	–65°C	15°C
Diurnal temperature range	–89 to –31°C	10 to 20°C
	(Viking 1 lander data)	(Standard atmosphere)
Others	Martian dust storms	

Fig. 3. *Some environmental data of present Mars and present Earth (source: Horneck, Gerda. The Microbial World and the Case for Mars, Planet. Space Sci. 48 (2000): 1053–1063).*

processes, indicated by volcanoes, canyons, and channels that resemble dry river beds. These extensive fluvial features, which were confirmed during several follow-on missions to Mars, were difficult to reconcile with any origin other than liquid water. They attest to a stable flow of water on Mars at some time in the past, and sporadically even in more recent times. In addition, Mars shows several similarities with the Earth: it has a similar diurnal rhythm and four seasons that alternate on both hemispheres (Figure 3). Hence, Mars is considered the key target for the search of life beyond the Earth.

6.2.2. Past Mars

There is geological and mineralogical evidence of the presence of liquid water on early Mars, during the first few 100 million years. With the Mars Exploration Rover (MER) mission of NASA, the story of water on the red planet has been further unravelled. At the landing site of the rover Opportunity, distinct layering in some rocks showed that water once flowed there on the surface of Mars, leaving ripple-like curves in the outcrop rocks. Bead-like objects, the so-called "blueberries", turned out to be rich in hematite, a mineral that requires water to form. The detection of sodium chloride which only forms when water has been present is another indication of liquid surface water in the past of Mars.[232] More insight into the history of water on Mars has been obtained from measurements with the OMEGA (Observatoire pour la Mineralogie, l'Eau, les Glaces et l'Activité) instrument of the Mars Express mission of ESA.[233] The global mineralogical data further support the supposition of an aqueous environment on early Mars, i.e., during the Noachian period (up to 3.5 billion years ago) indicated by the formation of clay minerals. This period, probably alkaline, was followed by a more acidic one in the Hesperian period (up to 1.8 Ga ago), as indicated by the massive occurrence of hydrated sulphate minerals. These conditions were driven by extensive outgassing of volatiles coupled with a rapid drop in atmospheric pressure. Liquid water was probably still present during transient and local events, such as volcanic activity, impact release, or the melting of ice deposits. Therefore, by analogy to the early Archean biosphere on Earth, where fossil microorganisms as old as 3.5 billion years have been detected, an early Martian biosphere could be postulated with habitats and microenvironments similar to those on the early Earth. However, in contrast to the Earth, Mars became gradually more and more dry and cold, thereby loosing the capability of sustaining conditions clement for life. Tackling questions whether Martian life ever arose on the planet and, if so, whether it still might exist in some protected niches plays a central role in the scientific exploration of Mars.

6.2.3. Present Mars

During the last Amazonian period (up to present) Mars has been cold and dry, as also indicated by the presence of anhydrous ferric oxides[233]. From the global neutron mapping of the Mars Odyssey mission, the present distribution of water in the shallow subsurface was divided in four types of regions: (i) regions with dry soil with a water content of about 2 wt.%; (ii) northern permafrost regions with a high content of water ice (up to 53 wt.% of water); (iii) southern permafrost regions with high content of water ice (>60 wt.% of water) covered by a dry layer of regolith; and (iv) regions with water-rich soil at moderate latitudes (about 10 wt.% of water) covered by a dry layer of soil. These water-rich regions are well separated from the Martian atmosphere by a rather thick layer of desiccated regolith. Therefore, it was supposed that they were formed a long time ago when the climate allowed liquid water on the surface.[234]

More information has been provided by the on-board measurements of the spacecraft Mars Express and Mars Reconnaissance Orbiter, currently orbiting Mars. Prominent results of the current Mars Express mission are the detection of deep underground water-ice at the South Pole by the Mars Advanced Radar for Subsurface and Ionospheric Sounding (MARSIS) instrument estimating a total volume of 1.6×10^6 km^3 of water, which is equivalent to a present global water layer of about 11 m,[235] the discovery of large-scale explosive volcanism on recent Mars (about 350 Ma ago),[236] indications of relatively young volcanic activities in the north polar ice regions,[237] and the global distribution of anhydrous and hydrated minerals.

However, compared to the conditions on present Earth, the surface of Mars is extremely hostile to life as we know it (Figure 3). This is mainly due to the high influx of cosmic radiation, the wide spectral range of solar UV radiation including UV-C, the low pressure and the low temperatures and their large diurnal fluctuations. On Mars, CO_2 is the major absorber of the short wavelength UV radiation. As a consequence, UV radiation at wavelengths >200 nm reaches the surface of Mars. Hence, although the solar constant for Mars is only 43% of that for Earth, the fraction of biologically effective UV radiation reaching the surface of Mars is by far greater than that reaching the surface of the present Earth. The second feature one should note is the attenuation of solar radiation by dust particles occasionally dispersed in the atmosphere.

Therefore, if life once started on Mars, the gradual decreasing pressure and temperature as well as the intense solar UV radiation might have forced the emerging biota to retreat to some protective oases, where it might persist even today. Potential oases, to which putative life on Mars might have withdrawn are inferred from terrestrial analogues, such as deep subsurface rocks inhabited by cryptoendolithic microbial communities, the polar ice caps and permafrost regions, submarine or sub-ice hydrothermal vents or other hydrothermal areas in connection with volcanic activities, or endoevaporites, i.e., microbial communities that live in salt crystals, e.g., halite or gypsum.[238]

The exploration program of the European Space Agency (ESA) foresees as the next step the ExoMars mission that uses a rover with high autonomy, equipped with the analytical capacity to select suitable drilling site or exposed vertical stratigraphy to find signs of extinct or extant life.[239] To do this requires the development of an efficient drilling system and the use of the corresponding sample analysis suite in the underground exploration of selected sites. In addition, the habitability of these regions will be explored by in situ measurements of the climate, radiation environment and surface and subsurface chemistry in dry and wet state. It is important to understand the mechanisms of the strong oxidative processes present on the surface of Mars which have been investigated by the Viking experiments.[240]

In the endeavour to search for signatures of life on Mars, it is assumed that several robotic missions will have to precede any human landing on Mars, which is currently planned sometime after 2030. Finally human and robotic missions should be complementary and so that astrobiology could then immensely benefit from human presence on Mars.

Human exploratory missions to the Moon or Mars are part of the future planning of ESA and NASA and are widely considered the next logical step of peaceful cooperation in space on a global scale. Besides the human desire to extend the window of habitability, human exploratory missions are driven by several aspects of science, technology, culture and economy. The continued study of the evolution of Mars may even contribute to the understanding of the evolution of the whole solar system. Human spaceflight may be important in this research context, since a number of items have been identified where the action of astronauts in situ could be beneficial to reach the scientific goals, such as site identification by local analysis, sample acquisition at these sites, sampling and – if a laboratory is available on Mars – supervision of sample analysis, etc. Meeting the scientific objectives of a Mars mission will require autonomous and smart tools such as intelligent sample selection and collection systems on a very high level of automation and robotics. As soon as human travellers are involved, the need for integrated advanced sensing systems will become obvious such as for bio-diagnostics, medical treatment, and environmental monitoring and control. Furthermore, the development and test of technologies for in situ resource utilization (for producing propellant from atmospheric CO_2 or from water ice, but also for life support purposes) may turn out to be a powerful technology stimulus.

Human exploratory missions to the Moon and to Mars also have the potential of promoting peaceful cooperation on a global scale. The International Space Station (ISS) is the first example of an international cooperative venture for the joint development, operation, and utilization of a permanent space habitat in low Earth orbit (LEO), involving nearly all space-faring nations. Hence, with the ISS, a new era of peaceful international cooperation in space has started. Major potential partners are the U.S., Russia, Japan, Europe and Canada, with the U.S. taking the leading role. China has just recently joined the nations involved in human spaceflight. Lessons learned from this experience gained with ISS may help all nations to become engaged in large future international space projects, creating harmony through common scientific endeavour.

Such long-duration missions beyond LEO will add a new dimension to human space flight, concerning the distance of travel, the radiation environment, the gravity levels, the duration of the mission, and the level of confinement and isolation the crew will be exposed to. This will raise the significance of several health issues. The areas involved include as follows: (i) radiation risks, especially from solar particle events, (ii) very long 0-gravity levels during interplanetary transfers, followed by very high gravity levels at Mars arrival (up to 6 g during aero-capture and landing) with severe consequences on the human body, (iii) almost no

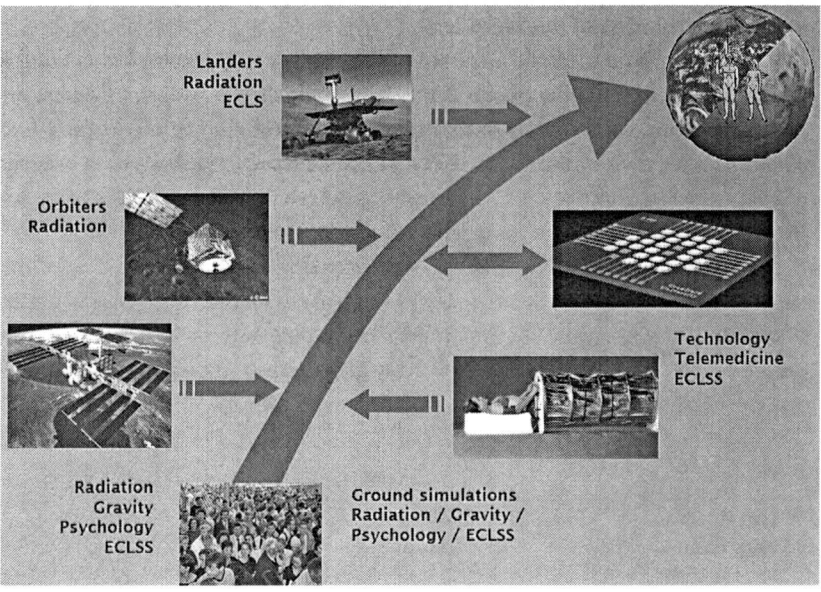

Fig. 4. *Roadmap in human health issues for ESA's exploration programme (source: Horneck, Gerda, and 15 co-authors. HUMEX, a Study on the Survivability and Adaptation of Humans to Long-Duration Exploratory Missions, ESA SP 1264, Noordwijk: ESA-ESTEC, 2003).*

mission abort or fast return capability, (iv) psychological issues which pertain to crew size, composition and corresponding education.[241] Substantial research and development activities are required in order to provide the basic information for appropriate integrated risk management, including efficient countermeasures and tailored advanced life support systems, as outlined in a roadmap for future European activities in life sciences in preparation of human exploratory missions (Figure 4), recommended in the HUMEX study of ESA.[242]

Furthermore, the import of internal and external microorganisms inevitably accompanying any human mission to Mars, or brought purposely to Mars as part of a bio-regenerative life support system needs careful consideration with regard to planetary protection issues.[243]

6.2.4. Future Mars

Wherever and whenever life has established itself, it has shown a natural instinct to expand ultimately the area from which its resources are drawn. The driving forces for migration to new settlements can be endogenous (e.g., hormones) or exogenous (e.g., deprivation of food or space, overcrowding by other species, unfavourable environmental conditions, light stimulus). Migration has certainly played a role in the evolution of our biosphere.

If migration is an intrinsic attribute of life, then it may inspire humankind to reach past the boundaries of the Earth in pursuit of the Moon, of Mars, and beyond. Natural catastrophes or anthropogenic global change might even force people to leave their home grounds and to invade remote, perhaps even extraterrestrial domains. Curiosity and the spirit of research as well as the urge to explore the unknown will be powerful drivers of human space exploration.

However, the human settlement of Mars may also threaten to have a significant environmental impact on the integrity of the planet. To safeguard sites of natural beauty, but also scientifically important sites, a planetary park system has been proposed to be established well ahead of any long-term exploration, exploitation or settlement of Mars.[244]

---

[231] Cockell, Charles S. "Habitability". Complete Course in Astrobiology. Horneck, Gerda and Rettberg, Petra, eds. Weinheim: Wiley–VCH, 2007. pp. 151–177.

[232] Squyres, Steve W., et al. "In Situ Evidence for an Ancient Aqueous Environment at Meridiani Planum, Mars". Science 306 (2004): 1709–1714.

[233] Bibring, Jean-Pierre, et al. "Global Mineralogical and Aqueous Mars History Derived from OMEGA/Mars Express Data". Science 312 (2006): 400–404.

[234] Tokano, Tetsuya, ed. Water on Mars and Life. Berlin; New York: Springer, 2005.

[235] Plaut, Jeffrey J., et al. "Subsurface Radar Sounding of the South Polar Layered Deposits of Mars". Science 316 (2007): 92–95.

[236] Hauber, Ernst, et al. "Discovery of a Flank Caldera and Very Young Glacial Activity at Hecates Tholus, Mars". Nature 434 (2005): 356–361.

[237] Neukum, Gerhard. "Recent and Episodic Volcanic and Glacial Activity on Mars Revealed by the High Resolution Stereo Camera". Nature 432 (2004): 971–979.

[238] Horneck, Gerda. "The Microbial World and the Case for Mars, Planet". Space Science 48 (2000): 1053–1063.

[239] Vago, Jorge L. and Gerhard, Kminek. Putting Together an Exobiology Mission: The ExoMars Example. Complete Course in Astrobiology. Horneck, Gerda and Rettberg, Petra, eds. Weinheim: Wiley-VCH, 2007. pp. 321–351.

[240] Klein, Harold P. "Did Viking Discover Life on Mars?" Origins of Life and Evolution of Biospheres 29 (1999): 625–631.

[241] Horneck, Gerda, et al. "HUMEX, a Study on the Survivability and Adaptation of Humans to Long-Duration Exploratory Missions, Part II: Missions to Mars". Advances in Space Research 38 (2006): 752–759.

[242] Horneck, Gerda, et al. "HUMEX, a Study on the Survivability and Adaptation of Humans to Long-Duration Exploratory Missions". ESA SP 1264, Noordwijk: ESA-ESTEC, 2003.

[243] Horneck, Gerda, et al. "Astrobiology Exploratory Missions and Planetary Protection Requirements". Complete Course in Astrobiology. Horneck, Gerda and Rettberg, Petra, eds. Weinheim: Wiley-VCH, 2007. pp. 353–387.

[244] Cockell, Charles S. and Horneck, Gerda. "A Planetary Park System for Mars". Space Policy 20 (2004): 291–295, Cockell, Charles S. and Horneck Gerda. "Planetary Parks – Formulating a Wilderness Policy for Planetary Bodies". Space Policy 22 (2006): 256–261.

6.3 Philosophical and religious implications of extraterrestrial intelligent life

Paolo Musso

If any evidence of extraterrestrial intelligent life were ever to be discovered, what would the likely reactions of our society be, and how should we manage such a discovery? The answer depends largely on the following question: What are the implications of such a discovery for our religious beliefs? This issue is the core subject of this chapter. Two basic assumptions are made: (1) There will not be direct contact with extra-terrestrials (ETs) or with intelligent probes, only indirect contact via electromagnetic waves, as in the "classic" SETI scenario.[245] (2) The contact will take place in the "near" future (i.e., in a few decades), which very likely means within a range of some hundreds of light years, due to the limits of our present technology.

6.3.1. The big issue

The first and foremost question is: Are we alone? Of course, if contact were actually achieved, the answer to that question would be: No! This is obvious, but has a not-so-obvious implication. In fact, it would mean that not only we are not alone, but also intelligent life is *very* common in the universe. In fact, under the assumptions stated above, the contact would be achieved in any case "in our cosmic backyard". Now, the probability of two highly improbable independent events occurring within such a small space is practically zero. Thus, if two intelligent civilizations were to actually exist so close to each other, then they would be (*very* likely) highly *probable* events. This fact has deep implications for many important issues: philosophical and religious, which are strictly related to each other. Furthermore, since there is no reason why the first civilization we will eventually meet should have something special,[246] it is very likely (though not absolutely sure) that it will be typical. So, this will allow us to believe that at least the most general statements we make about such a civilization have good chances of holding universally, also with respect to all other possible civilizations, including the ones we will never be able to contact.

6.3.2. Philosophical issues

The contact with another civilization would have implications with respect to many classical philosophical problems. Here are some of the most important ones:

6.3.2.1. Chance vs. necessity

Learning that intelligent life is common in the universe would not solve, of itself, the eternal debate about the role of chance and necessity in evolution. However, it would imply that at least its outcome (i.e., life and intelligence) should not be considered casual any longer. This would mean that either evolution is based on some (still unknown) deterministic mechanisms or there are some deterministic constraints forcing the random processes to follow a limited set of paths.

6.3.2.2. Are science and reason universal?

This point has been often doubted by contemporary epistemology, which is almost completely relativistic, and this attitude also has supporters within the SETI community. However, SETI is a very difficult task even assuming that ETs' technology is identical to ours. If their technology were to be completely or even partially different, contact would never be achieved. Thus, *if* contact were achieved, it would mean that their technology is actually identical, or at least very similar.[247] But this is possible if – and only if – science is universal, that is, if – and only if – science is actually the discovery of the objective laws of nature, and not a mere byproduct of societal dynamics, as the relativists claim. However, this would be possible if – and only if – reason is universal, too. That is, if – and only if – reason is a capability to know and understand reality in itself, and not if reason is a mere by-product of neuronal dynamics (and thus fully dependent on a particular evolutionary path).

6.3.2.3. The mind–body problem

There are many different theories about the mind–body problem (materialism, idealism, Cartesian dualism, Aristotelian hylemorphism, functionalism, and so on). But which one is the correct one? Of course, a full answer could never come from science alone, due to the nature of the question itself. Nonetheless, new interesting insights might come with the answer to the question whether intelligence is necessarily related to: (1) carbon chemistry; (2) DNA; (3) a humanoid shape.

6.3.2.4. What values would ETs have?

Of course, ETs may have ethical and religious values very different from ours. However, it is very unlikely that we would not even be able to understand each other. Indeed, to be intelligent beings means also to be moral beings, since it implies by definition: (1) to be able to imagine different possible futures; (2) to be able to choose among them rationally. Now, this is all we need to be moral beings, too. Indeed, a moral choice is nothing but a rational choice with consequences for the others. Thus, intelligent beings are also unavoidably moral (or immoral) beings (immorality being a moral category). In philosophical terms, morality is a structure (i.e., a constitutive dimension) of reason.

That is not all. ETs should be able to understand our religious ideas, too. ETs may have revealed or traditional religions or not, indeed, but surely they must have developed at least metaphysics. Metaphysics (i.e., the rational discussion about God) is essentially based on: (1) the problem of the First Cause of the world; (2) the problem of the ultimate foundation of ethics. Thus, beings who are both scientific and moral (as ETs would be) cannot avoid having answered in some way the two problems stated above, even if their answer were a negative solution, or even to declare the whole matter meaningless. Also these anti-metaphysical positions are, in fact, metaphysical of themselves, since (as Aristotle already noticed) the philosophical category to which a proposition belongs depends on what type of questions it wants to answer. Thus, scientific and moral beings (as ETs) are also unavoidably metaphysical beings. In other words, metaphysics is a structure (i.e., a constitutive dimension) of reason (which theologians also called "religious sense", to distinguish it from the particular metaphysical and/or from the particular religious ideas generated basing on it).

Now a problem arises. If, indeed, we will be able to understand ET's values, what should our position be? A very common mistake must be avoided in this context. Many believe that, since ETs are likely to be much more advanced than us from a scientific point of view,[248] their moral and religious beliefs would also be more advanced, so that it would be obvious that we should adopt them without any discussion. But to be more *advanced* does not necessarily imply to be more *moral*. Any modern dictatorship (such as Nazi Germany, the Soviet Union, Chinese communism, and so on) is much more advanced than any ancient society, but surely *much* less moral than most of them. The standard objection is that this may be true for the short term, but, over the long term, scientific advancement surely needs a certain degree of spontaneous cooperation. This is very likely to be true, but cooperation can also arise for non-ethical reasons (for example, intelligent, well-planned egoism or fear of a global catastrophe or self-destruction). Such a development could produce very efficient, but not necessarily very moral systems.

Furthermore, despite a very widespread opinion, to be more advanced does not imply even to be more intelligent. First, there is no evidence that human beings living on Earth at present are more intelligent than our pre-scientific ancestors, but just the contrary. We know very well, in fact, that any child from a community still living (from a technological point of view) in the Neolithic era (as – say – Australian or Amazon aboriginals), would have the same chances as any other child to become a good scientist if attending a modern school. Furthermore, from an evolutionistic point of view, it would simply be nonsense to believe that our intelligence would have been able to increase significantly not just from Galileo's times, but even from the Neolithic era. In other words, our entire scientific and technological progress has happened without implying *any* change in our intelligence. Cultural evolution indeed, does not work in the same way as the biological one, since it is essentially a matter of accumulation, that concerns society as a whole, and not individuals: "We are dwarfs on the shoulders of giants," and for this reason we can see farther, but our sight is not better than that of our carriers.[249]

Thus, ET's moral and metaphysical beliefs, as well as their traditional or revealed religions (if any), should be examined very carefully (provided that we are able to solve the linguistic problem and communicate this type of concepts, of course), but should not be considered *a priori* "better" than ours. On the contrary, they should be discussed rationally, just as it happens (or *should* happen) on Earth every day with ours. Furthermore, we should feel free to tell them *our* beliefs, too, without any "inferiority complex". ETs would be our partners in a dialogue, not our teachers!

6.3.3. Religious issues

In this part, I will only discuss the problems related to our Western religious tradition, i.e., Christian religion. There are three good reasons for this approach. (1) It would not be correct to speak in name of other people. (2) At present, SETI is only being done in some Western countries. (3) Christianity is usually considered the most anthropocentric religion and, as such, the one most likely to have problems, should ETs be detected.[250]

6.3.3.1. Christo-centric, not anthropocentric

The first point to be clarified in this context is the myth that Christian religion would be anthropocentric. This is based mainly on the (supposed) merger of Christian theology and geocentric Aristotelian philosophy based on the Ptolemaic system as regards its astronomic claims. Not only is today's situation rather different, but even Middle Age geo-centrism was not anthropocentric. Even in Aristotelian philosophy, the centrality of Earth was only geographical, not moral or metaphysical. On the contrary, the sub-lunar world was considered the realm of imperfection. This aspect was further accentuated in the Middle Age by the Christian doctrine of original sin. In fact, from this perspective, things were not very different from what they are today. The truth is that Christian religion is Christo-centric, not anthropocentric. What makes Earth a special place is the Redemption made by Jesus Christ. Now, even though Christ was a man, He was also God. This makes the situation very different from simple anthropocentrism, since it has also a cosmic meaning, although it is indubitable that humankind has a special (but not necessarily unique) role in the Christian view.

6.3.3.2. The plurality of worlds

Another commonplace to be ruled out is that the idea of a plurality of worlds has been condemned as heretical by the Catholic Church. This is completely false, even in the case of Giordano Bruno, who was condemned for other heretical theories, but not for this one. His theory of a plurality of inhabited worlds was *suspected* to be heretical, was discussed in depth during the trial, but finally was not condemned. There is full consensus in recent historiography in this matter.[251] Furthermore, we should not forget that Bruno did not properly speak of a *plurality* of worlds, but of an *infinity*, which is quite different, and likely to give raise to a number of paradoxical consequences, not only from a Christian point of view. This is not meant to justify the fact that he was burnt, of course, but only to put the question in the right context. On the contrary, in 1277 (that is three centuries before Bruno's trial) Stephan Tempier, then Archbishop of Paris, condemned as heretical the opposite proposition maintained by Averroist philosophers, i.e., "That the First Cause cannot create a plurality of worlds" ("Quod Prima Causa non posset plura munda facere"), since it denied God's omnipotence.

6.3.3.3. ET's existence

Nevertheless, what about the positive existence of extraterrestrials, i.e., the possibility that God has actually created them? Until today, there has been no systematic theological discussion, and, in particular, no official statement made by the Catholic Church about this subject. However, three Popes have expressed their opinions, briefly and informally, about it, one against (Zachary I), and two in favor (Pius XII, John XXIII). Most theologians (both Catholic and Protestant) who have discussed this topic are in favor of ET's existence, but we must be aware that they are very few. At present, the most common position within the Christian community is to wait and see.

6.3.3.4. How dramatic?

At this point, we would like to try to imagine some possible consequences that are likely in the case of a contact with an extraterrestrial civilization. First of all, under our assumption that only an indirect contact via electromagnetic waves will be achieved, no dramatic change (panic, revolutions, loss of authority, and so on) should be expected, at least over the short period, despite the fact that catastrophic predictions have been made very often. But why? Without a direct contact, ETs cannot be seen as a threat to us from the material point of view. The same applies from the spiritual perspective, because the existence of ETs has turned out to be, of itself, fully compatible with Christian faith, while their religious views, if disturbing, could be questioned or simply ignored, and also with good reason, as stated above. Furthermore, as a matter of fact, many people, including many Christians, already believe in ETs' existence without any problem. Thus, the most likely hypothesis is that, apart from an initial period of general excitation, our daily life on Earth would go on in more or less the same way.

A more difficult issue from a Christian point of view would be to establish the right place of ETs in the history of salvation (and thus our correct relationships with them), due to Christ's unique role. This would seem to imply that ETs should also have the possibility of a direct contact with believers in Christ, or, at least, of receiving the Gospel. In this sense, it is interesting to notice that the impossibility of wide communication with ETs, a factor that is usually considered to make the contact less dramatic, might be *more* dramatic from a Christian point of view. Without a convincing solution to this question, problems might arise for Christianity, but only over the long run. First, the impossibility of dialogue should be proven – and this is very likely to take centuries, if not millennia, unless contact happens within the range of few light years, which is highly improbable. Second, people usually want to know about their own salvation and that of their relatives and friends, but nothing more. One may blame such an attitude as selfish and petty (although Christ did not tell us to love humankind, but one's neighbour – that is much more difficult), however the matter stands thus. Therefore, it is very unlikely that many people could ever worry about ETs' salvation and the complex theological issues related to it. If – and only if – the attempt to solve this problem were to lead Christian theology to dramatic self-contradictions, with consequences also on the principles guiding *our* life, then a progressive process of loss of credibility might start, which, after many decades or even centuries, could be destructive for Christianity. But, once again, this is not very likely. Some solutions have already been proposed. For example, there may be a plurality of Incarnations (one for each civilization). It is even possible that ETs do not need any salvation, because are not affected by the original sin. But also a more "conservative" solution is possible.

6.3.3.5. "The Kingdom is like a seed . . . "

After all, Jesus was not born in the capital of the Roman Empire, but in one of its most neglected and disparaged places (even by the Hebrews themselves: "Could anything good ever come from Bethlehem?"). This fact would imply a sort of harmony, first of all aesthetical, in discovering that He was also born in one of the most neglected and disparaged places (even by Terrestrials themselves . . .) of the Galactic Club. And, as scientists know very well, aesthetical harmony is always a powerful sign that we are on the right track. In more theological terms, according to the Christian view God's "method" seems to be to choose a small group of people, working like a "seed". Even though the seed must develop and grow throughout history, the fully developed "tree" will never be visible in this world, but only in the other, at the end of time. As a matter of fact, even most of the intelligent beings that have lived on Earth have never heard of Christ. Nonetheless, He also died for each of them. So, the same might apply to ETs, meaning that neither a direct contact nor communication of the Gospel would be necessary.

6.3.4. How should we manage such a discovery?

In summary, ETs are expected to firstly have a more advanced science, but one that is not substantially different from ours; secondly, to have possibly different moral and religious values, but to be able to understand ours and discuss it rationally; thirdly, to be challenging but acceptable for Christian religion; and fourthly, not to endanger our present social system. On this basis, we will try to outline some recommendations on how to manage the discovery of an extraterrestrial signal.[252]

6.3.4.1. Transparent communication

Since it cannot pose a danger to us, the discovery of an extraterrestrial message as well as its content should be communicated to the public without any delay and in a fully transparent way.

6.3.4.2. Avoiding religious conflicts

As contact with ETs should not cause true religious conflicts, it would be important to avoid creating false ones. Here are some suggestions:

(1) We should avoid defining in advance the "meaning" of the contact and its implications for humankind in any official document or communication, recognizing that there is a wide range of different possibilities and, above all, that the *real* meaning of the contact, both for humankind and for any particular culture, will be decided, ultimately, by each human being living on Earth at that time and later.

(2) We should avoid the "common values" trap, accepting and respecting all existing cultural differences, recognizing that whatever divides us may be not less important than what unites us. I agree with the common view that "Earth should speak with one voice," but if – and only if – this voice is more similar to a polyphonic chorus than to a "solo".

(3) We should think about a suitable reply in advance, taking in consideration not only scientific communication, but also the cultural aspects, which is a much more difficult task. In fact, a common reply may well be the only way to avoid an uncontrolled, wild proliferation of private replies: indeed, finding a signal by searching the sky at random is very difficult, but sending a signal to a precise target is very easy. With the exception of some international workshops promoted by the SETI Institute starting in 2001, nothing has been or is being done about the latter issue. My own personal proposal[253] is to construct the message according to a "federal" style. The first part should be a joint undertaking and contain the "dictionary" (i.e., the definition of the meaning of the symbols used in the message), some general information about us and our planet from a scientific point of view, and an introduction written on behalf of humankind describing our common values. The second part should be written – using the same code as the first one – by the delegates of the main terrestrial cultural traditions, both religious and non-religious, each one presenting their own beliefs.

6.3.5. Appendix: Managing ET's technology

The only possible threat in the detection of an extraterrestrial message would be if it were to contain a detailed description of alien technology. This does not seem very likely to happen, but, if it does, serious troubles might arise, since the massive introduction of an alien technology could be very destabilizing for our society.

In principle, such a decision should not be taken by governments and scientists alone, but also by the common people. However, this might be very hard to put into practice. If we were to receive only a self-proclaiming message (as, e.g., a radio-carrier) or a low-information one, it would be conceivable (even though not so easy) to organize a worldwide democratic consultation after an in-depth debate. In the case of a positive outcome, we could ask ETs to share their knowledge with us. In the opposite case, we could ask them not to do so and to refrain from considering further messages asking the opposite. But what if a wide exposition of alien technology were already encoded in the earliest message? Should we stop the decoding process until the terrestrial population has decided? Then, in the case of a negative result, should we simply destroy this part of the message? If so, how could we ever be sure that it has actually be done, and that nobody has kept a secret copy to gain an incommensurable advantage over all the other people in the future? How could we ever be sure that we are not losing information essential to decoding the rest of the message? The scenario might even be even worse. For example, the message could be very powerful and transmitted continuously, again and again, many times over, so that anyone with an antenna, even a very small one, would be able to receive it. In this case, of course, we would forgo the problem of making any decision at all.

However, things might also be *better*. Indeed, despite the almost universally widespread opinion, the fact that ETs would almost definitely be much *older* than us does not imply that they would be much more *advanced*. On the contrary, the opposite is much more likely. In fact, this equivalence is based on the assumption that technological progress is an endless process. Nonetheless, without substantial advancements in science we can only improve the existing techniques until the full exploitation of their possibilities, as established by the laws of nature which they are based on. Thus, endless technological progress implies endless scientific progress. Yet, no phenomenon in nature can grow indefinitely. On the contrary, phenomena usually follow a logistic curve, showing initial quasi-exponential growth followed by progressive deceleration until a state of saturation is reached. More specifically, the ultimate aim of science is to discover the fundamental laws of nature. Now, if science cannot succeed in this, then it will have to stop somewhere before reaching its goal. But if it *can* succeed, then it will discover *all* the laws and then it will have to stop. Thus, *in any case* science (and therefore technology, too) has to stop, soon or later.

Now, the decision of whether or not to ask ETs to share their knowledge with us will depend on: (*a*) the gap in scientific progress between ETs and us; (*b*) the distance between ETs and us. Indeed, should *a* (expressed in years) be smaller or equal than $2b$ (expressed in light years), their reply would reach us only when it would have become useless, since, by then, we should have already reached the same degree of advancement. Now, whereas *b* is very easy to measure when we detect ETs' message, *a* is very difficult. However, when considering the progress humankind has achieved in the last four centuries, it seems rather strange that grasping the most fundamental laws of nature will take another 1 billion years. A statistical analysis of the main discoveries and inventions that I made 5 years ago[254] shows a couple of classical bell-shaped patterns following roughly the same trend, where the curve of technology is delayed by about 25 years. This is a rather remarkable value because it corresponds to about one generation, that is, exactly the time one may reasonably expect to be needed to "translate" scientific discoveries into technological gears. Even though the inception of science was four centuries ago, most of the progress achieved to date started only two centuries ago. Moreover, if I am right, progress already entered its descendent phase some decades ago. Therefore, scientific progress on Earth may substantially stop after no more than two or three centuries, and technological progress just a little later. Therefore, even a civilization 1 billion years older than us should not be more than two or three centuries more advanced than us. Although further studies are needed, at this stage I feel confident enough to conclude that we are virtually certain that we will never have to face any dramatic choice, since the minimal distance of the Earth from the nearest civilization (if any) is very likely greater than 100 light years, so that $2b \geq 200$ years $\geq a$.

---

[245] SETI is the acronym for Search for Extra-Terrestrial Intelligence. In favor of this scenario there are not only technological reasons, but also the so-called "Fermi Paradox", which postulates that, if interstellar travels are possible and ETs exist, they should already be here. Musso, Paolo. What the Fermi Paradox Tells us about the Dangers of Active SETI. 2006. (Talk given at the 57th International Astronautical Congress (IAC), Valencia, 4–8 October 2006, presently available at www.filosofiadellascienza.it).

[246] This is the so-called "Copernican Principle" (sometimes also named "Principle of Mediocrity").

[247] ET's science might be, of course, less or (very likely) more *advanced* than ours, and in this sense "different." But *only* in this sense. It could not be different in the sense that ETs have a different knowledge of the same objects.

[248] Since our civilization was "born yesterday" on the cosmic scale. ETs might be *much* older than us, indeed, even by *more than one billion years*. Norris, Ray. "How Old is ET?" When SETI Succeeds: The Impact of High-Information Contact. Tough, Allen, ed. Washington DC: Foundation for the Future, 2000. pp. 103–105.

[249] Musso, Paolo. "How Advanced Is ET?" Life in the Universe: From the Miller Experiment to the Search for life on Other Worlds. Seckbach, Joseph, et al. eds. Dordrecht; Boston and London: Kluwer Academic Publishers, 2005. pp. 335–337.

[250] Vakoch, Douglas. "Roman Catholic Views of Extraterrestrial Intelligence: Anticipating the Future by Examining the Past". When SETI Succeeds: The Impact of High-Information Contact. Tough, Allen, ed. Bellevue, Washington: Foundation for the Future, 2000. pp. 165–174.

[251] Ciliberto, Michele. Giordano Bruno. Bari; Rome: Laterza, 1990.

[252] Tarter, Jill, and Michaud, Michael, eds. "SETI Post-Detection Protocol". Special Issue of Acta Astronautica 21 (1990).

[253] Musso, Paolo. "Wide Cultural Communication in Interstellar Messages". Bioastronomy 2002: Life Among the Stars. Norris, Ray P. and Stootman, Frank H, eds. Ann Arbor, Michigan. International Astronomical Union, 2004. pp. 511–513.

[254] Musso, Paolo. "How Advanced Is ET?" Life in the Universe: From the Miller Experiment to the Search for life on Other Worlds. Seckbach, Joseph, et al. eds. Dordrecht; Boston and London: Kluwer Academic Publishers, 2005. pp. 335–337.

6.4   ET culture

Debbora Battaglia

This chapter invites the reader to enter the outerspaces of extraterrestrial culture, as a realm of social inquiry. Where this journey leads is perhaps unexpected, especially for the discourse of aliens and Unidentified Flying Objects (UFOs). For while we might expect to engage fields of exotic Otherness – of technomarvels and bizarre entities, epic enterprises, and terrors unrecognizable in their "structures of feeling" – we find ourselves instead in the presence of an extraterrestrial uncannily familiar and concrete: visible and heard in realms of mass media and popular culture, and in the iconic signatures of new religious movements, on the one hand, while making visible the limits of official culture authority structures – the limits of "the powers that be" – on the other hand. From this perspective, Extra Terrestrial (ET) culture does not allow us to avoid the question of what counts as knowledge and truth and for whom, right here on Terra.

Especially in insecure times, alien knowledge communities have much to teach us about ourselves, anthropologically speaking. Their voices in this volume raise fundamental questions that include Who are we? Where did we come from? Are we alone as a species? What do we make of our human differences? In raising these questions, we unsettle the boundaries of Us and Other, human and non-human things and entities, in no uncertain terms.

6.4.1. The alien de-exoticized

Extraterrestrial practices are about possible world making. Accordingly, interdisciplinary "galaxies of discourse" are not merely a luxury, but necessary to any new vision of humans in outer space. The idea of the extraterrestrial quite naturally articulates fields of social science, science and technology studies, linguistics, popular and expressive culture, religion and spirituality, and social and intellectual history. The net effect is to unsettle the boundaries of science, magic, and religion, and to query common knowledge about sites of technological and human agency. Accordingly, there is a creative happiness to the enterprise. In pragmatic terms, we must make creative leaps of faith out of our disciplinary comfort zones, if we are to realize our potential to connect with different entities in space. This kind of abdicative reasoning, as the philosopher George Sanders Peirce referred to it, is the very condition of hypothesis-making. In short, it is important to admit a *de-exoticized alien* into our realms of critical thinking, and too, into the light of our most searching disciplinary questions.

As we come face-to-face with the *ET – effect* and in particular, with the evidence of how it performs a kind of magic upon everyday phenomena and draws people together into new configurations of community within its aura, our boundaries of self and society are productively unsettled. For the simple reason that the idea of off-planet life forces us to take a reflexive, planetized view of the material consequences of unequal relations of power, of our cultural imaginaries, our own history, even what it means to see the world "from the native point of view" is opened to interrogation. For many who believe in the possibility of intelligent life and unknown technologies of contact, *science fiction is science fact*, identity is never a given, and the boundaries of inside and outside are always open to interrogation. Particularly in the networked spaces that we inhabit today, when information of unknown origin enters our living spaces as a matter of course, and often unbidden, this book does not allow us to forget that the diversely inhabited worlds which contribute to life's ambiguous messages and contingencies must be taken seriously.

We can turn on our televisions to the National Geographic Channel and hear that astronomers and astrophysicists and other real scientists believe that extraterrestrial life is not preposterous – but probable. While it is not the purpose of this chapter to interrogate the truth-value of such claims or of the documents circulating as evidence in spheres of public culture, this book shows that that there is much to be learned at this historical moment by turning anthropology's apparatus of visibility to sites of extraterrestrial culture, and to bear witness to the leaps of faith of those who embrace the idea of the alien, in terror and in hope.

6.4.2. Galaxies of space discourse

Anthropology's ethnographic method of participant-observation carries with it a certain value for the context of the "the field". Classically this notion calls up images of fieldwork and written representations of unified "societies" and "cultures," and their "institutions" and "customs" in remote places. And indeed, all of these categorical units of analysis lend themselves readily to cross-cultural comparison. In late nineteenth century's Age of Empire, they aided the young field of anthropology in defining itself as a coherent discipline, distinctively positioned to gain new knowledge of "Us" and "Others".

Having come a long way from such "othering" binaries practices, contemporary anthropology finds new purpose in interrogating the differences and the gaps – the creative tensions – between cultural orders and complex daily realities, and in the actions people take to connect, and disconnect, across them. This is true to the nth degree for engaging the social in culture-specific terms of outer space, which abides, and thrives, in open questions, and articulates an astonishing range of social issues "from the Earth native's point of view" (as anthropologist Susan Lepselter puts this so well.[255] Indeed, a striking feature of ET culture and its *galaxies of discourse*, is the extent to which conventionally distinct fields of knowledge cross-connect, collide, or pass through one another, under its influence. For "insider" and more detached researchers alike, these galaxies reconfigure the facts and relationships of particular human lives in their present day and, often unbeknownst to actors, in terms recalling the science religiosity of earlier times. Outer spaces, in its multiplicity of genres, are in this sense deeply cultural and explicitly historical. But more to the point, they are *invitational sites* that call us to the horizons of subjects' inner spaces, and demand that we hone apparatuses of hospitality; that we rethink the terms of reference of space diplomacy.

Thus, the chapters in this book, with their multiple approaches to alien/UFO experiences, communities, networks, and science and technology, are congruent with topologies of spacetime, multi-sited from the start. Specifically for anthropology, the validity of being "elsewhere" is a given of the subjectivities we explore. So, too, is the validity of being "elsewhen," whether in the "lost time" of an abduction experience, or "in the beginning" of a religious philosophy of alien creationism, or on a pilgrimage to the history-making *faith-sites* of contact and exchange – to Roswell, New Mexico, where *everyone knows* that the first "flying saucer" crashed in 1947, or to the "mother of all crop circles," created in 2001 by unidentified artists in the grain fields of southern England. In fact, the diversity of faithsites – published dialogues and narratives of contact and abduction and dwelling among off-planet entities, ufological archives, sacred inscriptions, and UFO theme museums, talk radio airspace, archaeological sites, and so forth – poses a methodological challenge to outer space research. Since our object of study is discursively fluid, the problem of discerning local "models of and for" the enterprise (as the anthropologist Clifford Geertz famously recognized) cannot be a simple one.

6.4.3. Modeling an ET diplomacy

However, it is an element of Bruno Latour's network theory[256] that intrigued me in relation to ET culture. As anthropologist Karen Sykes observes, Latour's "passion for scientific practice [which] verges upon religiosity... *deepens the intimacy* between the ideal and the material by repeatedly cross-cutting their separation".[257] This quality, and in particular the "passion" and "intimacy" of the engagement, are characteristics of ET culture – igniting the quest for relating desire and reason, and I would add, prompting also a certain humility.

As opposed to the scientific gaze that would seek to "*get rid* of all the filters one brings to an object of study in order 'to see things are they are,'" Latour sees the project as "'giving an opportunity' to phenomena that, in other settings, would not be 'given a chance' [to appear]"[258] by way of the alternative path of the *proposition:* "offers made by an entity to relate to another under a certain perspective".[259] He continues, contrasting the value of the proposition to that of (linguistic and optical) metaphors, which take interpretative biases as interfering with accuracy: "The more activity there is, and the more intermediaries there are, the *better* the chance to articulate meaningful propositions".[260] Safford, commenting, draws the point clearly: "Each entity is forced to pay attention to the other, and, in so doing, both diverge from their customary paths to venture into territory which, although it appears foreign from each of their unique vantage points, nonetheless belongs to an interdependent existence".[261]

That this interdependence must be taken on faith is important. Anthropological inquiry calls forth the subjectivities in the proposition, and offers a new approach to a practical extraterrestrial imaginary of interdependence – what I shall refer to as an *anthropological model of visits.*[262] Drawn from the image of the World Wide Web's standing invitation to "only connect," this model opens diplomacy to the subjunctive mood, *as if* neither scale nor power asymmetries in fact mattered. After all, space is not only about states, and neither is it (only) about culture. It follows that a historically situated (as opposed to an overarching, universalistic) space law, and clear guidelines for legal negotiations that recognize networked exchanges and their documents as an artifact of localized networks, are a crucial element of any model of visits.

Further, this requires that entities on both sides of an exchange take positions of what might be termed courageous vulnerability vis-à-vis the other; both needing "to pay attention to the other, and, in so doing, both diverg[ing] from their customary paths to venture into territory which, although it appears foreign from each of their unique vantage points, nonetheless belongs to an interdependent existence".[263] Overall, then, ET diplomacy is hospitable to, and returns us to, the notion of a *de-exoticized alien-to-alien exchange.* Less simple than it may appear, the idea of "the visit" challenges us to reevaluate our attachments to, and detachments from place. Thus, as we trace the decontexualizations and recontextualizations of extraterrestrial events, entities, and objects, it is impossible to deny the vulnerability and contingency of the social contexts subjects inhabit. On the one hand, we cannot set aside the problematic of departures and disconnects, while we focus on the work that anthropology is more deeply invested in of how connections form across differences. On the other hand, as migration studies have long understood, neither can we give place of privilege to progressive narratives and "rites of passage" without at least considering how passages and migration generally speaking might be imagined as circular. From this perspective, the question becomes how "home" is experienced upon return, and whether one really can go home again.

Observe what Russian cosmonaut Valentin Lebedev wrote in his diary in 1982, on the day before he and his fellow cosmonaut, Anatoly Berezovoy, would return to Earth after seven months in space on board the spacecraft Solyut-7: "We are anxious; who knows why." And what about? His dangerous reentry? The prospect of landing in a remote area of Russia during a snowstorm and waiting for over half an hour in freezing conditions to be discovered as rescue helicopters crashed nearby? (These things actually happened.) No. He is wondering, "What's it like down there? We're no longer accustomed to life on the ground. Our lives are attuned to this small island in space, and suddenly here we come, back to the Big World! We don't feel comfortable with the idea".[264] And how worrying if they had been comfortable with the idea!

Whether we are focused on extraterrestrials (aliens) or alien technology (ufology); on channels of communication or saucerian visions, each homes to event-sites and event-time right here on Terra. The idea of humans in outer space presents an opportunity to consider carefully the conceptual tools that we take along with us, and as well to acknowledge that (as astronauts designate the greatest danger in space) we deepen our understanding of the weather that threatens our mission there – in the all-too-human sense of patterns of violence and hegemonic political agendas that subvert our human capacity for creativity and connection.

6.4.4. The ET effect upon the social

The chapters in this book demonstrate an exploratory project in two registers: involving human subjects and alien entities at sites of faith in outer space – not least, interdisciplinary conferences such as the one in which most of them originated. Partly because of this structural affinity, the two sets of relations can productively engage one another in their departures from more usual grounds of authority for recognizing modalities of the human project. And since the metaphysical is an illuminating but not essential component of this reflexive exchange, they can together raise a common, and profound, question regarding who claims the status of "host" and who "visitor" – as it were (paraphrasing the French philosopher, Jacques Derrida), who dares to say welcome at these sites. We do so in anticipation that anthropological questions will yield new knowledge that supplements and destabilizes prior knowledge, acting back upon the field, and productively revealing new gaps and insufficiencies there, and new itineraries of discourse.

It is imperative to hear this: the *ET-effect* cannot be dismissed as science fantasy before we know what precisely of social and material consequence to a heterogeneous life on Earth we are actually dismissing; what the "extra" in Extra Terrestrial is doing for and to us. Questions of transparency in authority, authorship, and authorization will figure centrally in any enlightened space policy, alongside questions about sources of power and access to knowledge. In a positive vein, the idea of an alien knowledge source can inspire bold efforts of translation across cultural differences, carrying with it the promise of new horizons of social exchange. In a negative vein, alien powers can call up a common nemesis: the opaque and inaccessible "powers that be" which, as noted by political theorist Jodi Dean,[265] guard access to the domains of secret knowledge.

Recent anthropology has raised questions about this domain's invisibility or transparency, respectively, and has produced groundbreaking collections that emphasize the tyranny of the one[266] or of the other,[267] respectively. Refusing to write the culture concept out of the picture, these collections focus much-needed attention on operations of modernity and postmodernity that would seem actually to warrant anxiety, from the subject's point of view. In the main, this anxiety is focused on a Truth that is "out there," to cite the signature theme of conspiracy studies, but hidden and controlled by an absolutely powerful few – to disastrous effect for future life on Earth. Whether such authorities are the scientific establishment's "vigilante skeptics",[268] or moral police within world religions, the consequences for subjects who challenge them are in some sense shared by all Earth dwellers. From this perspective, it is not inappropriate to construct their, and our, concerns in terms of boundary maintenance, creating more openness to the resources that corporeal bodies, the body politic, and celestial bodies need in order to flourish.

In this chapter I have sketched some thoughts toward a new use for anthropology as a reflexive, historically and socially situated futurology for planetized persons and their intellectual, spiritual, and popular culture communities. Here, the idea of the extraterrestrial shows itself as an exemplary site for exploring insiders' voices in outer spaces, and too, for exploring the fullest reach and range of anthropological critical interrogation, theory, and methods: to productively destabilize prior knowledge of the field. I have tried to expose otherness (though not a sense of the foreign) as fraudulent to the discourse of the alien, that we might reveal human differences of extraterrestrial consciousness and social practice in more appropriately familiar terms. The collage of interdisciplinary fields of this book bears witness to the complexity of interdiscursivity, generally speaking, and in strict accord with the entities it considers, reveals the Vienna Vision as a project both terrifying, and hopeful beyond belief.

[255] Lepselter, Susan. "From the Earth Native's Point of View: The Earth, the Extraterrestrial, and the Natural Ground of Home". Public Culture 9 (1997): 197–208.

[256] Latour, Bruno. The Well-Articulated Primatology: Reflections of a Fellow Traveler. Primate Encounters: Models of Science, Gender, and Society. Strum, Shirley C. and Fedigan, Linda M, eds. Chicago: University of Chicago Press, 2000. pp. 358–381.

[257] Sykes, Karen. "My Aim Is True: Postnostalgic Reflections on the Future of Anthropological Science." American Anthropologist 30 (2003): 156–168.

[258] Latour, Bruno. The Well-Articulated Primatology: Reflections of a Fellow Traveler. Primate Encounters: Models of Science, Gender, and Society. Strum, Shirley C. and Fedigan, Linda M, eds. Chicago: University of Chicago Press, 2000. p. 368.

[259] Ibid. p. 372.

[260] Ibid. p. 375.

[261] Safford, Barbara Maria. Visual Analogy: Consciousness as the Art of Connecting. Cambridge, MA: MIT Press, 1999.

[262] Battaglia, Debbora, ed. "ET Culture: Anthropology in Outer Spaces." Durham, NC: Duke University Press, 2005.

[263] Cited in Safford, Barbara Maria. Visual Analogy: Consciousness as the Art of Connecting. Cambridge, MA: MIT Press, 1999. p. 183.

[264] Lebedev, Valentin. Diary of a Cosmonaut: 211 Days in Space. Trans. Luba Diangar. New York: Bantam Books, [1988] 1990. p. 268.

[265] Dean, Jodi. Aliens in America: Conspiracy Cultures from Outerspace to Cyberspace. Ithaca: Cornell University Press, 1998.

[266] Marcus, George. Paranoia Within Reason. Chicago: University of Chicago Press, 1999.

West, Harry G. and Sanders, Todd, eds. Transparency and Conspiracy: Ethnographies of Suspicion in the New World Order. Durham: Duke University Press, 2003.

[267] Strathern, Marilyn. Introduction: New Accountabilities: In Audit Cultures: Anthropological Studies in Accountability, Ethics and the Academy. Strathern, Marilyn, ed. London: Routledge, 2000. pp. 1–18.

[268] Collins, Harry M. and Pinch, Trevor J. Frames of Meaning: The Social Construction of Extraordinary Science. London: Routledge, 1982.

CHAPTER 7

THE VIENNA VISION
ON HUMANS IN OUTER SPACE

Humans in Outer Space

The way forward for the next 50 years

Space age has reached its 50th anniversary. Development of robotic exploration to distant planets and bodies across the solar system, as well as pioneering human space exploration in Earth orbit and the Moon, paved the way for ambitious long-term space exploration. Europe has always played a significant role in the endeavours of humankind to explore other worlds and to understand the Universe in which we live. Today, space exploration goes far beyond a merely technological endeavour, as its further development will have a tremendous social, cultural and economic impact. Space activities are now entering an era where the contribution of the humanities – history, philosophy, anthropology, the arts as well as the social sciences – political science, economics and law – will become crucial for the future of space exploration. Now that the awareness for the societal complexity of activities in space is growing internationally, it is vital that Europe, with a stronghold in natural sciences as well as its identity firmly rooted in the humanities and the social sciences, grasps the opportunity to involve their specific knowledge(s) in the long-term planning of exploration undertakings.

The next generation may be given the opportunity to explore new places and discover new worlds. Those adventures will be driven by the human desire of quest for knowledge and human curiosity. They will provide a main opportunity for equitable international cooperation. Humans divided on Earth will hopefully unite in space as citizens of one planet.

Interdisciplinary Odysseys

The European Science Foundation (ESF) has organized the first comprehensive trans disciplinary dialogue on humans in outer space. This dialogue goes further than regarding humans as better than – robot tools for exploration. It investigates the human quest for odysseys beyond Earth's atmosphere and reflects on the implications of the findings of extraterrestrial life. The inherent human curiosity for exploring the unknown is at the heart of this dialogue, and is addressed through collaboration between the ESF Standing Committee for the Humanities (SCH) and the ESF European Space Sciences Committee (ESSC), in cooperation with the European Space Agency (ESA) and the European Space

229

Policy Institute (ESPI) in Vienna. Recently the "Athens declaration" enabled by the ESSC established a scientific framework for defining Europe's exploration programme.

The **Vienna Vision on Humans in Outer Space** was developed at the "Humans in Outer Space" conference, held in Vienna on 11–12 October 2007 locally organised by ESPI with the support of the Austrian Ministry for Transport, Innovation and Technology (BMVIT). This vision provides a European perspective in identifying the relevant needs and interests linked with space exploration. It is presented to several European and international fora, in order to make it a useful element for the position-finding and decision-making process.

First Odyssey

Humans in Earth orbit

What effect does it have?

Home. Earth is a fragile oasis in the vastness of the solar system and it needs to be protected from natural and man-made threats. Once in space, humans are no longer just citizens of individual countries, but also of the planet Earth caring for its overall global sustainability.

Progress. Human space flight is a major source of innovation. It can benefit societies around the world with a variety of technological spin-offs and scientific research possibilities; it has even been argued that it could help to overcome the limits for growth on several levels. Through endeavours such as the International Space Station new partnerships are built, which can cultivate international cooperation in a spirit of friendship and mutual understanding.

Technology. Humans increasingly rely on technological advancements in their everyday lives. Relationship between humans and machines will reach new dimensions, and in the process may make it necessary to readjust our notion of 'humanity'. Space applications can have a positive impact on the quality of life on Earth and eventually beyond. Through television and internet everyone can virtually experience space flight or the vistas of planetary surfaces. In the near future space tourism may no longer be a dream but become a possibility to those interested.

Law. The legal framework for space activities needs to be further developed in a way that cultivates peaceful uses of outer space and equal rights for all humankind. Human rights will also have to be considered, as new moral challenges will face humanity.

Second Odyssey

Humans in space exploration

What effects will it have?

Humanity. In the new era of technological advancements, the human factor is essential. Without human presence in space, spaceflight and exploration will lack an important dimension. Global cooperative endeavours will allow fostering the further development of collaboration among peoples, societies and cultures.

Discovery. Space exploration allows for discovery in two ways: it makes it possible to search for specific things, i.e. new energy resources; but it also opens up the opportunity to follow the thrust of scientific and cultural curiosity. The latter is one of the most inspiring traits of humankind since the beginning of its history and it should lead again to incredible discoveries.

Culture. Space exploration is a challenging, cooperative endeavour that offers opportunities to further strengthen European ties and define European values and priorities. The identity of Europe is constituted by its specific cultural approach towards both scientific and moral issues, and it will be this angle which will influence societal development as well as serve as inspiration for the younger generations.

Rights. Through space exploration, new partnerships will form. This will call for a proper legal framework serving to peacefully regulate issues such as space traffic management. Furthermore planetary protection needs to be elaborated with international partners concerning forward and backward contamination, and Europe must play an influential role in that context.

Third Odyssey

Humans migrating the Earth

How will it affect human thought?

Habitat. Driven by curiosity and in order to extend opportunities, humans may eventually search for settlements outside our planet. What is unimaginable today may become necessary in the future. The first child to be born in space will mark the dawn of a true space generation.

Encounters. Humans should be open to the idea of possible encounters with other forms of life in outer space, either through the discovery of life in the solar system (extinct or extant), or through the reception of extraterrestrial radio signals. A new era will begin should humans realize that they are not alone in the universe. Such discovery may likely cause the development of a new collective identity for humanity.

Belief systems. What people believe in, and how such beliefs are structured, has a strong binding force on societies, on Earth and eventually beyond. Human belief systems, whether religious or secular, change in the context of new living environments, and in contact with other forms of life and societies. As the merely technological or political approach will no longer be sufficient in dealing with such contacts, the humanities and the social sciences will gain in importance.

Adapting. Past encounters that took place on Earth show that human beings did eventually adapt to unforeseeable realities, although often at very great costs. While the first effects of an encounter between humans and extraterrestrial life are unpredictable, humans need to be aware that they will be held morally, economically and politically accountable for their choices.

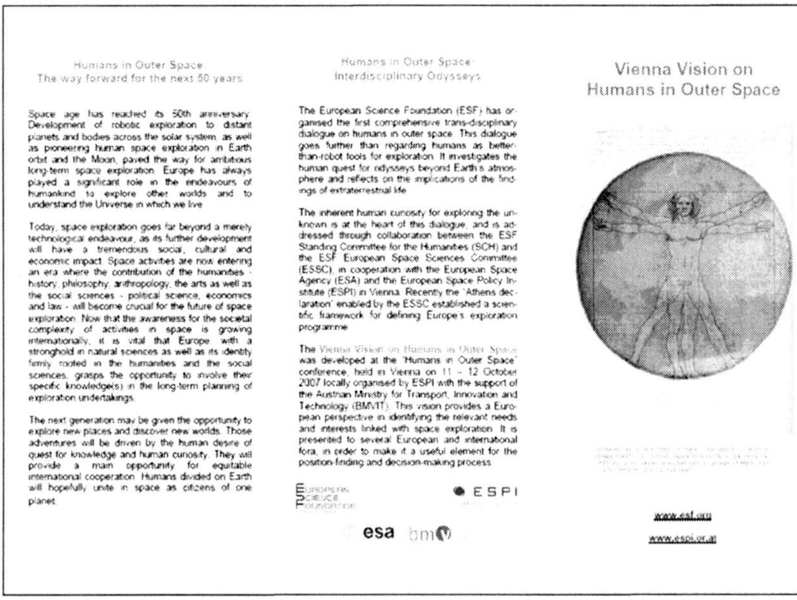

Humans in Outer Space:
The way forward for the next 50 years

Space age has reached its 50th anniversary. Development of robotic exploration to distant planets and bodies across the solar system, as well as pioneering human space exploration in Earth orbit and the Moon, paved the way for ambitious long-term space exploration. Europe has always played a significant role in the endeavours of humankind to explore other worlds and to understand the Universe in which we live.

Today, space exploration goes far beyond a merely technological endeavour, as its further development will have a tremendous social, cultural and economic impact. Space activities are now entering an era where the contribution of the humanities - history, philosophy, anthropology, the arts as well as the social sciences - political science, economics and law - will become crucial for the future of space exploration. Now that the awareness for the societal complexity of activities in space is growing internationally, it is vital that Europe, with a stronghold in natural sciences as well as its identity firmly rooted in the humanities and the social sciences, grasps the opportunity to involve their specific knowledge(s) in the long-term planning of exploration undertakings.

The next generation may be given the opportunity to explore new places and discover new worlds. Those adventures will be driven by the human desire of quest for knowledge and human curiosity. They will provide a main opportunity for equitable international cooperation. Humans divided on Earth will hopefully unite in space as citizens of one planet.

Humans in Outer Space:
Interdisciplinary Odysseys

The European Science Foundation (ESF) has organised the first comprehensive trans-disciplinary dialogue on humans in outer space. This dialogue goes further than regarding humans as better-than-robot tools for exploration. It investigates the human quest for odysseys beyond Earth's atmosphere and reflects on the implications of the findings of extraterrestrial life.

The inherent human curiosity for exploring the unknown is at the heart of this dialogue, and is addressed through collaboration between the ESF Standing Committee for the Humanities (SCH) and the ESF European Space Sciences Committee (ESSC), in cooperation with the European Space Agency (ESA) and the European Space Policy Institute (ESPI) in Vienna. Recently the 'Athens declaration' enabled by the ESSC established a scientific framework for defining Europe's exploration programme.

The Vienna Vision on Humans in Outer Space was developed at the 'Humans in Outer Space' conference, held in Vienna on 11 – 12 October 2007 locally organised by ESPI with the support of the Austrian Ministry for Transport, Innovation and Technology (BMVIT). This vision provides a European perspective in identifying the relevant needs and interests linked with space exploration. It is presented to several European and international fora, in order to make it a useful element for the position-finding and decision-making process.

ESPI

esa bm

Vienna Vision on
Humans in Outer Space

www.esf.org

www.espi.or.at

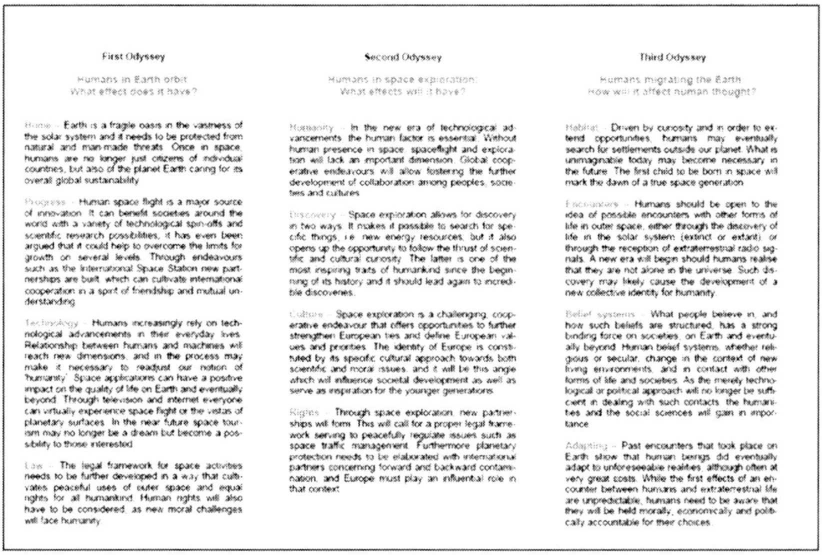

First Odyssey

Humans in Earth orbit:
What effect does it have?

Home – Earth is a fragile oasis in the vastness of the solar system and it needs to be protected from natural and man-made threats. Once in space, humans are no longer just citizens of individual countries, but also of the planet Earth caring for its overall global sustainability.

Progress – Human space flight is a major source of innovation. It can benefit societies around the world with a variety of technological spin-offs and scientific research possibilities, it has even been argued that it could help to overcome the limits for growth on several levels. Through endeavours such as the International Space Station new partnerships are built, which can cultivate international cooperation in a spirit of friendship and mutual understanding.

Technology – Humans increasingly rely on technological advancements in their everyday lives. Relationship between humans and machines will reach new dimensions, and in the process may make it necessary to readjust our notion of 'humanity'. Space applications can have a positive impact on the quality of life on Earth and eventually beyond. Through television and internet everyone can virtually experience space flight or the vistas of planetary surfaces. In the near future space tourism may no longer be a dream but become a possibility to those interested.

Law – The legal framework for space activities needs to be further developed in a way that cultivates peaceful uses of outer space and equal rights for all humankind. Human rights will also have to be considered, as new moral challenges will face humanity.

Second Odyssey

Humans in space exploration:
What effects will it have?

Humanity – In the new era of technological advancements, the human factor is essential. Without human presence in space, spaceflight and exploration will lack an important dimension. Global cooperative endeavours will allow fostering the further development of collaboration among peoples, societies and cultures.

Discovery – Space exploration allows for discovery in two ways. It makes it possible to search for specific things, i.e. new energy resources, but it also opens up the opportunity to follow the thrust of scientific and cultural curiosity. The latter is one of the most inspiring traits of humankind since the beginning of its history and it should lead again to incredible discoveries.

Culture – Space exploration is a challenging, cooperative endeavour that offers opportunities to further strengthen European ties and define European values and priorities. The identity of Europe is constituted by its specific cultural approach towards both scientific and moral issues, and it will be this angle which will influence societal development as well as serve as inspiration for the younger generations.

Rights – Through space exploration, new partnerships will form. This will call for a proper legal framework serving to peacefully regulate issues such as space traffic management. Furthermore planetary protection needs to be elaborated with international partners concerning forward and backward contamination, and Europe must play an influential role in that context.

Third Odyssey

Humans migrating the Earth:
How will it affect human thought?

Habitat – Driven by curiosity and in order to extend opportunities, humans may eventually search for settlements outside our planet. What is unimaginable today may become necessary in the future. The first child to be born in space will mark the dawn of a true space generation.

Encounters – Humans should be open to the idea of possible encounters with other forms of life in outer space, either through the discovery of life in the solar system (extinct or extant), or through the reception of extraterrestrial radio signals. A new era will begin should humans realise that they are not alone in the universe. Such discovery may likely cause the development of a new collective identity for humanity.

Belief systems – What people believe in, and how such beliefs are structured, has a strong binding force on societies, on Earth and eventually beyond. Human belief systems, whether religious or secular, change in the context of new living environments, and in contact with other forms of life and societies. As the merely technological or political approach will no longer be sufficient in dealing with such contacts, the humanities and the social sciences will gain in importance.

Adapting – Past encounters that took place on Earth show that human beings did eventually adapt to unforeseeable realities, although often at very great costs. While the first effects of an encounter between humans and extraterrestrial life are unpredictable, humans need to be aware that they will be held morally, economically and politically accountable for their choices.

Fig. 1. *Flyer 'The Vienna Vision on Humans in Outer Space'.*

About the authors

Jacques Arnould is an agricultural Engineer. He obtained a Ph.D. in History of Sciences and Ph.D. in Theology. He is taking an active interest in the interrelation between sciences, cultures and religions with a particular interest for two set of themes: the first related to the life sciences and his evolution; the second related to space conquest. To the first he devoted several works and publications on the historical and theological aspect. To the second, he is the French Space Agency (CNES) expert in charge of ethical, social and cultural aspect of space activities.

Thomas Ballhausen is head of the Department for Studies and Advanced Research at Filmarchiv Austria in Vienna, lecturer at the University of Vienna and guest lecturer at the University of Applied Arts Vienna. Thomas Ballhausen has published several books and articles in the fields of film history, media theory and popular culture. He is co-editor of the book series *Materialien zur österreichischen Filmgeschichte* and editor of the book series *exquisite corpse – Schriften zu Ästhetik, Intermedialität und Moderne*. Recent publications include *Psyche im Kino. Sigmund Freud und der Film* (co-editor with G. Krenn and L. Marinelli, verlag filmarchiv austria, 2006), *Die Unversöhnten* (Skarabeaus, 2007) and *Delirium und Ekstase. Die Aktualität des Monströsen* (Milena, 2008).

Debbora Battaglia is a socio-cultural anthropologist specializing in comparative issues of identity and personhood in realms of public culture. Her most recent research with alien/UFO believers focuses on questions of science, religion, and self in modernity. Her books include the anthologies *E.T. Culture: Anthropology in Outerspaces* (Duke University Press) and *Rhetorics of Self-Making* (University of California Press), and she is the author of *On the Bones of the Serpent: Person, Memory and Mortality in Sabarl Island* (University of Chicago Press). In addition she has received major awards from the John Simon Guggenheim Foundation and the National Endowment for the Humanities. She lives in Massachusetts, where she is Professor of Anthropology at Mount Holyoke College.

Wolfgang Baumjohann is currently Director of the Institut für Weltraumforschung of the Austrian Academy of Sciences in Graz (Aurtria). He has also been recently elected Vice President of ESA's Science Programme Committee (SPS). His fields of research include space plasma physics and planetary magnetism. Wolfgang Baumjohann's work has been published widely of which the most cited publication is: *W. Baumjohann, G. Paschmann, C. A. Cattell: Average plasma properties in the central plasma sheet. J. Geophys. Res., 94, 6597, 1989 (275 Citations* *in SCI*). In addition he has participated in various projects, inter alia, as principal Invesitgator (Project Lead) Scandinavian Magnetometer Array, Equator-S Magnetometer, Equator-S Data Center, German Cluster Data Center, BepiColombo/MMO Magnetometer. Amongst many other memberships, Wolfgang is member of International Academy of Astronautics and most recently Member of ESA's Earth Science Advisory Committee.

Ulrike M. Bohlmann is a German lawyer working for the European Space Agency since 2002. She has covered legal issues of ESA's Science and Earth Observation Programmes. Her main tasks consist in the preparation, negotiation, and submission for approval by the Agency's delegate bodies of international agreements. Before joining ESA, Ulrike Bohlmann worked at the Institute of Air and Space Law of the University of Cologne, where she also earned her doctoral degree magna cum laude with a thesis on commercial space activities and industrial property protection.

Amedeo Cesta is a senior research scientist at ISTC-CNR, where he has founded and currently leads the Planning and Scheduling Team (PST – http://pst.istc.cnr. it/). He received a master degree in Electronic Engineering and a Ph.D. in Computer Science at the University of Rome "La Sapienza" in 1983 and 1992, respectively. He has conducted research in several Artificial Intelligence areas like Multi-Agent Systems, Intelligent Human-Computer Interaction, Planning & Scheduling and always pursued the synthesis of innovative Cognitive Systems. His work focuses on the integration of planning and scheduling in software architectures, the use of constraint programming for specialized tasks such as temporal reasoning, the synthesis of planning and scheduling heuristics, the interactive solution of complex planning and scheduling problems. His work in Artificial Intelligence also emphasizes the real-world aspects of automated reasoning, in particular focusing on research topics like robust and flexible planning and scheduling, execution monitoring and mixed-initiative systems.

Luca Codignola (DL Rome 1970, M.A. Toronto 1974, DLitt hon. Saint-Mary's 2003), is Head of the Institute of History of Mediterranean Europe of the Italian National Research Council. He is also Professor of North American History at the University of Genoa, and Adjunct Professor at Saint Mary's University. His main field of research is the Roman Catholic Church in the North Atlantic area in the early modern era. He has also written on the early European expansion. His latest publications are Columbus and Other Navigators (2007); "Roman Catholic Conservatism in a New North Atlantic World, 1760–1829," William and Mary Quarterly (2007); and "The Holy See and the Conversion of the Aboriginal Peoples in North America, 1760–1830" (2008).

Gabriella Cortellessa is a research associate at ISTC-CNR since 2000. She received her Master degree in Computer Science Engineering at the University of Rome "La Sapienza" in 2001, with a thesis on HCI techniques applied to planning and scheduling systems. In April 2005 she defended her Ph.D. thesis in Cognitive Psychology at the same university, focusing on an experimental analysis of interactive problem solving features. Her Ph.D. program has been supported by a grant from ISTC-CNR on funding from Italian and European Space Agencies projects. In 2003 Gabriella Cortellessa has spent one year at Carnegie Mellon University as a visiting student scholar. Her research spans on Mixed Initiative Problem Solving, Automated Synthesis of Explanations, Experimental Methods for Evaluating Intelligent System features. She collaborates with both the Environmental Psychology Unit at ISTC-CNR and the Department of Cognitive Psychology of the University of Rome "La Sapienza". Gabriella's main current interest lies in the synthesis and robust evaluation of cognitive systems applications.

Alfred Worcester Crosby, Jr. (B.A. Harvard 1952, M.A. Boston University 1956, Ph.D. Boston University 1961), served in the U.S. Army (1952–1955) and is now Professor Emeritus of American Studies at the University of Texas, Austin, where he has taught since 1977. As a historian of science, culture, and the environment, in his many books he has touched upon and attempted to explain several key issues in world history. The Columbian Exchange (1972) and Ecological Imperialism (1986), devoted to the influence of disease and biological exchange in a transoceanic context, are regarded by historians as key historiographical turning points in the history of European expansion. His latest books are: Throwing Fire. Projectile Technology through History (2002) and Children of the Sun. A History of Humanity's Unappeasable Appetite for Energy (2006).

Monique van Donzel studied French Literature and Linguistics and General Linguistics at Leiden University, The Netherlands. She received her Ph.D. from the University of Amsterdam in 1999 in Phonetic Sciences. Her thesis investigated the use of prosody to realize and recognize the structure of spoken discourse in spontaneous speech in Dutch. She then joined the Netherlands Organisation for Scientific Research (NWO) and was involved in the evaluation of the NWO research institutes, before joining the NWO Council for the Humanities, where she was responsible for international affairs. As of January 2004 she is on secondment from NWO at the European Science Foundation (ESF), first as Science Officer for the Humanities and Social Sciences and since October 2005 as Head of the Humanities Unit and Senior Science Officer. Her main responsibilities include the development of strategic initiatives in the Humanities in Europe and the management of European networking activities funded through the ESF. In addition, she is senior scientific secretary to the Standing Committee for the Humanities (SCH).

Frans G. von der Dunk is currently Professor of Space Law at the University of Nebraska-Lincoln, College of Law. He was awarded the Distinguished Service Award of the International Institute of Space Law (IISL) of the International Astronautical Federation (IAF) in Vancouver, in October 2004, and the Social Science Award of the International Academy of Astronautics (IAA) in Valencia, in October 2006. He defended his dissertation on "Private Enterprise and Public Interest in the European 'Spacescape'" in 1998. Frans has written over 100 articles and published papers, has given some 100 presentations at international meetings and was visiting professor at some 20 foreign universities across the world on subjects of international and national space law and policy, international air law and public international law. He has (co-)organised some 20 international symposia, workshops and other events, and has been (co-)editor of a number of publications and proceedings. As of 2006, he is the Series Editor of 'Studies in Space Law', published by Brill. Finally, he has given a number of interviews to the international media on issues of space law and policy. Frans has served as adviser to numerous institutions and agencies. Much of his recent work furthermore focused on such topical issues as space tourism, the legal status of the Moon and other celestial bodies and the 'sale-of-lunar-estate hoax', and planetary protection. He is Director and Treasurer of the International Institute of Space Law (IISL) and has numerous memberships in international organisations a few amongst many others include: Member of the Board of the European Centre for Space Law (ECSL), Member for the Netherlands in the International Law Association's (ILA) Committee on Space Law and Member of the Editorial Board of 'Space Policy'.

Marcel Egli is head of the Space Biology Group, Swiss Federal Institute of Technology Zurich, Switzerland (ETH Zurich). He graduated at the University of Bern (M.D. – Ph.D.) before he spent several years as Postdoctoral Fellow at the University of Melbourne (Howard Florey Institute, Australia) and at the Florida State University (Department of Biology). Thereafter, he became Associate Faculty at the Florida State University. The appointment to lead the Space Biology Group brought him back to Switzerland. Marcel Egli has published various papers in the fields of neurobiology, neuro-endocrinology, and currently in space-biology. He is referee for papers submitted to the European Journal of Neuroscience, Endocrinology, Microgravity Science und Technology, and The Journal of Clinical Endocrinology & Metabolism. He is active member of the Society for Neuroscience, and the Society for Research on Biological Rhythms.

His current research interests include hypothalamic control of hormone secretion, and the influence of biological rhythms on immune function in humans. In particular the question how the internal biological clock is altered by microgravity. Furthermore, he fosters numerous collaborations with Universities of Applied Sciences in Switzerland to develop new technologies for space applications. He is Principal Investigator of several experiments that were selected by the European Space Agency (ESA) for a space flight. Marcel Egli is a regular guest lecturer at the University of Bern and the ESA Summer School for Astrobiology at Banyuls.

Olivier Francis is Professor of Physics at the University of Luxembourg. He obtained a B.Sc. in Physics, a Masters in Geophysics and a Ph.D. at the Catholic University of Louvain in Belgium. He was Visiting Scientist at the University of Colorado in Boulder (1997–1998), Associate Lecturer for the Open University, Newcastle upon Tyne (1994–1998), Research Assistant in the Groupe de Recherche en Géodésie Spatiale from the Centre National d'études Spatiales in Toulouse (1988–1993), and also Research Scientist at the Royal Observatory of Belgium (1988–2000). He served as Director of the European Center for Geodynamics and Seismology (ECGS) (2002–2006). He is a specialist in Earth and oceanic tides and in their interaction. He is involved in several projects to measure climatic change effects using geodetic and geophysical measurements. He is also active in metrology for accurate and precise measurements of gravity. Research involves fluid mechanics, numerical analysis and data processing. He is an author and co-author of 68 publications with 34 in peer-reviewed journals. He is member of the LESC Standing Committee and of the ESSC of the European Science Foundation.

Sven Grahn is a pioneer in Swedish space activities. He holds an M.Sc. from the School of Engineering Physics at the Royal Institute of Technology in Stockholm and a Ph.D. (h.c.) from the Institute of Technology in Stockholm. Sven has had leading roles in all Swedish Space Corporation's (SSC) satellite projects, and has been engaged in most other SSC projects. Since 2006 until present Sven is Senior Advisor to the Swedish Space Corporation and is involved in various other teaching and technical consultant activities. He has also been honored and achieved several awards for his achievements, inter alia, such as The Gold Medal from the Swedish Academy of Engineering Sciences (2002) and most recently the Gold medal for achievements in aeronautics and astronautics in memory of Swedish aviation pioneer Enoch Thulin from the Swedish Society of Aeronautics and Astronautics (2008).

Gerhard Haerendel received his Ph.D. (Dr. rer. nat.) in Physics from the University of Munich in 1963. In 1969 he became Fellow (Wissenschaftliches

Mitglied) of the Max-Planck-Institut für Physik und Astrophysik, and from 1972 to 2000 he was Director at the Max-Planck-Institut für extraterrestrische Physik (MPE). In 1987 he was appointed Honorarprofessor at the Technische Universität Braunschweig. He has been Visiting Professor at the University of Iowa in 1988 and at the University of California, Berkeley, in 2000. He was Dean of the International Space University in 1989 and from 2000 to 2005 Vice-President and founding Dean of the School of Engineering and Science of the International University Bremen (now Jacobs University). He remains affiliated with the Jacobs University as Distinguished Professor of Space Physics. From 1989 to 2001 he has been Vice President of the International Academy of Astronautics and from 1994 to 2002 President of the Committee on Space Research (COSPAR). From 2003 to 2007 he chaired the European Space Science Committee (ESSC). As of January 2007, he is chairman of ESA's advisory committee for Human Spaceflight, Microgravity and Exploration Programmes (ACHME). He has more than 40 years of experience in experimental and theoretical space plasma physics and plasma-astrophysics publishing more than 260 scientific papers. He is member of several professional societies and academies and recipient of several awards.

Gerda Horneck is former Head of the Radiation Biology Section and former Deputy Director of the Institute of Aerospace Medicine of the DLR. She has been involved in radiobiological and astrobiological space experiments on Spacelab, LDEF, EURECA, FOTON, MIR, ISS, and ExoMars. For her research Gerda Horneck was awarded several honors of ESA, NASA, DLR and the International Academy of Astronautics.

Ulrike Landfester is at present (since 2006) Member of the Standing Committee for the Humanities at the European Science Foundation, since 2004 Member of the Research Council of the Swiss National Foundation (Humanities and Social Sciences) and since 2003 full professor for German language and literature at the University of St. Gallen (Switzerland). She has obtained a Ph.D. at the University of Munich (Dissertation: Der Dichtung Schleier. Zur poetischen Funktion von Kleidung in Goethes Frühwerk – The Veil of Truth. The poetical function of clothing in Goethes early work), undertaken the study of German literature, English literature and Medieval literature at Albert-Ludwigs-Universität in Freiburg (Germany). Prior to that Ulrike has undertaken the study of Archeology, Egypteology and Early History at the aforementioned university in Freiburg. She has been the co-editor of the complete works of Rahel Levin Varnhagen (Edition Rahel Levin Varnhagen), University of Hamburg. Ulrike has also taught at several institutions; inter alia, University of Vienna (Austria), professor at the University of Konstanz, full professor at the University of Frankfurt am Main.

Stephan Lingner is Deputy Director of the "Europäische Akademie zur Erforschung von Folgen wissenschaftlich-technischer Entwicklungen Bad Neuenahr-Ahrweiler GmbH" since 2005. He is senior scientist with responsibility for
the Academy's research on technology and environmental assessment and coordinated several interdisciplinary projects thereupon. He is also managing editor of
the Springer journal *Poiesis & Praxis: International Journal of Ethics of Science and
Technology Assessment.* Before, he was research fellow at the German Aerospace
Center (DLR), Cologne, where he was responsible for systems analyses of new
options for spaceflight and space exploration. In his previous research he was
planetary scientist at Münster University, focussing on lunar rock analyses of
Apollo 14 samples. Stephan Lingner published numerous scientific articles and
reports on environmental protection, technology assessment and space science &
exploration and has been expert reviewer of the Intergovernmental Panel of
Climate Change (IPCC) and on other scientific projects and papers. He holds
a doctoral-degree in planetary chemistry and was a lecturer in "Ecology" at
Koblenz University of Applied Sciences from 2000 to 2007.

Agnieszka Lukaszczyk is, since September 2006, the Executive Officer of the
Space Generation Advisory Council (SGAC). Starting May 2008 she has also
been working as a space policy consultant at the Secure World Foundation (SWF).
In addition, during the period of September 2006 to June 2008, Agnieszka worked
at the European Space Policy Institute. She holds a Masters degree from the
American University's School of International Service in International Politics
and a Bachelor degree in Political Science form the University of Tennessee. She
also studied at the Université Catholique de Louvain in Brussels, Belgium, the
Jagiellonian University in Krakow, Poland and the World Trade Institute in
Berne, Switzerland. She interned at the Political Section of the Polish Embassy in
Washington, D.C., American Electronics Association in Brussels, European
Department of the Polish Senate in Warsaw and the Warsaw Business Journal.

James Muldoon is a graduate of Iona College (1957) and obtained an M.A. at
Boston College (1959). At Cornell University (1965) he studied with Professor
Brian Tierney, where he obtained a Ph.D. He then taught at St. Michael's College
in Vermont (1965–1970) and subsequently at the Camden (New Jersey) College
of Arts and Sciences of Rutgers University (1970–1998). He is now Professor
Emeritus at Rutgers, an adjunct instructor at the Rhode Island School of Design,
and a research fellow at the John Carter Brown Library on the campus of Brown
University. James has been a Fulbright Fellow at Trinity College of Cambridge
University where he studied with Professor Walter Ullmann (1963–1964). A
specialist in medieval legal and ecclesiastical history, James has a particular interest
in the relations between Christian and non-Christian societies in the medieval and

early modern world. This work has led to three books published by the University of Pennsylvania Press: *The Expansion of Europe: The First Phase (1977); Popes, Lawyers, and Infidels: The Church and the Non-Christian World 1250–1550 (1979); The Americas in the Spanish World Order: The Justification for Conquest in the Seventeenth Century (1994).* He also edited two volumes of essays on religious conversion, *Varieties of Religious Conversion in the Middle Ages (1997)* and *The Spiritual Conversion of the Americas (2004),* both published by the University Press of Florida. His other recent books are *Empire and Order: The Concept of Empire, 800–1800 (Macmillan, 1999) and Identity on the Medieval Irish Frontier (University Press of Florida, 2003).* He has also published more than 35 articles dealing with such topics as the treatment of Indians in colonial Massachusetts, the law of Christian marriage, Pope Alexander VI's division of the world between the Castilians and the Portuguese in 1493, the law of the sea, the origins of international law, and the legal issues associated with the crusades. Many of these articles have been reprinted in his *Canon Law, the Expansion of Europe, and World Order (Ashgate/Variorum, 1998).* At the moment, James is co-editing with Professor Felipe Fernández-Armesto a series of reprints dealing with the expansion of medieval Europe tentatively entitled *The Expansion of Latin Europe, 1000–1450.* He is also developing a series of articles of 18th century American perceptions of the Middle Ages and the implications of that knowledge for the American revolutionary generation.

Paolo Musso is Professor of Philosophy of Science at the University of Insubria of Varese and Visiting Professor of Epistemology at the Universidad Católica Sedes Sapientiae of Lima (Perù), after having been Professor of Philosophy of Nature at the Urbaniana Pontifical University and at the Pontifical University of the Holy Cross, both in Roma. At the end of 1997 he started working on Bioastronomy and, especially, SETI, focusing on its linguistic and philosophical problems, giving presentations (some published) about this topics at many international meetings. He was also *Rapporteur* of the SETI Session at the *International Astronautical Congress* (IAC) of Bremen 2003. At present he is full member of the *SETI Permanent Study Group* of the *International Academy of Astronautics* (IAA), member of the *European Exo/Astrobiology Network Association* (EANA), consultant of the *Observatory on Complementary Medicines* of the University of Verona and member of the Scientific Committee of the Italian scientific magazine *Emmeciquadro.* So far he has published about 30 articles and 4 books: *Rom Harré e il problema del realismo scientifico (Rom Harré and the problem of scientific realism),* Angeli, Milano, 1993, *Filosofia del caos (Philosophy of chaos),* Angeli, Milano, 1997, *Convivere con la bomba. Dalla bioetica alla biopolitica (Living with the Bomb. From Bioethics to Biopolitics),* ESI, Napoli, 2002, and *Forme dell'epistemologia* *contemporanea*, Urbaniana University Press, Roma, 2004. He has also collaborated with the Italian national newspapers *Avvenire* and *La Repubblica*.

Claude Nicollier has been for nearly 30 years a European Space Agency (ESA) astronaut. He graduated from the University of Lausanne, Switzerland in 1970 (Bachelor of Science in physics) and the University of Geneva in 1975 (Master of Science in astrophysics). He also graduated as a Swiss Air Force pilot in 1966, an airline pilot in 1974, and a test pilot in 1988 (Empire Test Pilot's School, Boscombe Down, United Kingdom). He was a member of the first group of ESA astronauts selected in 1978. He joined Group 9 of NASA astronauts in 1980 for Space Shuttle training at the Johnson Space Center, Houston, Texas, where he has been stationed until September 2005. His technical assignments in Houston have included Space Shuttle flight software verification in the Shuttle Avionics Integration Laboratory (SAIL), development of Tethered Satellite System (TSS) retrieval techniques, Remote Manipulator System (RMS) and International Space Station (ISS) robotics support. From 1996 to 1998, he was Head of the Astronaut Office Robotics Branch. From 2000 on, he was a member of the Astronaut Office EVA (Extravehicular Activity) Branch. During his assignment in Houston, he also maintained an active duty status within the Swiss Air Force with a rank as Captain. In October 2005, he was assigned to the European Astronaut Center in Cologne, Germany, to work on different projects within the ESA Human Space Program, including preparation of the next ESA astronauts selection, and operational support, from the ground, of the "Astrolab" mission, first long-duration stay of an ESA astronaut onboard ISS, from July to December, 2006. He retired from ESA on 31 March 2007, and is currently professor at the Federal Institute of Technology in Lausanne, where he teaches a course on "Space Technology and Operations", and provides assistance to students on various space related projects. He has been a crewmember on four Space Shuttle flights, STS-46 in 1992 (EURECA deployment and first test of TSS), STS-61 in 1993 (first servicing mission of the Hubble Space Telescope), STS-75 in 1996 (second flight of TSS, and USMP-3 microgravity investigations), and STS-103 in 1999 (third servicing mission of the Hubble Space Telescope). He has logged more than 1000 hours in space, including a spacewalk of 8 h 10 min duration to install new equipment on the Hubble Space Telescope on STS-103.

Angelo Oddi is a research scientist for the Institute of Cognitive Science and Technology of the Italian National Research Council (ISTC-CNR). He received a master degree in Electronic Engineering and a Ph.D. in Medical Computer Science at the University of Rome "La Sapienza" in 1993 and 1997, respectively. He has been a visiting scholar at the Intelligent Coordination and Logistics Laboratory of the Robotics Institute at Carnegie Mellon University in 1995–1996.

His work focuses on the application of Artificial Intelligence techniques for scheduling, automated planning and temporal reasoning. Regarding his professional activities, he has published more than 50 papers, both in journals and in peer-reviewed international conferences. He has a vast experience in the design of intelligent systems for real-world applications.

Gísli Pálsson (Ph.D., University of Manchester, 1982) is Professor of Anthropology at the University of Iceland and (formerly) the University of Oslo. He is Honorary Fellow of the Royal Anthropological Institute of Great Britain and Ireland and Associate Fellow at the Centre for Biomedicine & Society (CBAS), King's College, London. In 2000 he received the Rosenstiel Award in Oceanographic Science from the School of Marine and Atmospheric Science of the University of Miami. He has written over 100 articles in scientific journals and books and published and edited 15 academic books. Among his books are *Anthropology and the New Genetics* (2007), *Travelling Passions: The Hidden Life of Vilhjalmur Stefansson* (2005), *The Textual Life of Savants* (1995), *Nature and Society: Anthropological Perspectives* (1996, co-editor), and *Images of Contemporary Iceland* (1996, co-editor). Currently, Gisli's research focuses on the social implications of biotechnology, human tissue collections, genetic history, environmental change, and Arctic exploration. Gisli Pálsson has done anthropological fieldwork in Iceland, the Canadian Arctic, and The Republic of Cape Verde.

Nicolas Peter is since 2006 Research Fellow at the European Space Policy Institute (ESPI). He has been a Lockheed Martin Fellow for two years at the Space Policy Institute at the George Washington University (GWU) and has worked for the X Prize Foundation in Washington DC on future space prizes. Nicolas has also been a Trainee in the Science, Technology and Education Section of the European Union Delegation of the European Commission to the USA, as well as Teaching Associate for the International Space University's Master programme and a Faculty and Team Project Co-chair for the Summer Session Programme. Nicolas Peter has completed various research activities in Europe (France and Austria) North America (Canada and USA) and Asia-Pacific (Australia and Japan). His primary research interests are in space policy and international relations. Nicolas Peter has published and presented over 50 articles in peer-reviewed journals and international conferences related to space activities, particularly on space policy issues. He has also been invited to be rapporteur for sessions dealing with space policy affairs held in the framework of international space conferences in Canada, Spain and India. Nicolas holds a Bachelor of Geography from the Louis Pasteur University in Strasbourg, France. He holds also his first Masters Degree in Space Systems and Environment and second Masters Degree in Space Technology Applications from the Louis Pasteur University. Nicolas Peter is also a graduate from the International Space

University's Master in Space Studies programme and holds a Master of International Science and Technology Policy from GWU's Elliott School of International Affairs.

Kai-Uwe Schrogl is the Director of the European Space Policy Institute (ESPI) in Vienna, Austria since 2007. Before, he was Head Corporate Development and External Relations Department in the German Aerospace Center (DLR). In his previous career he worked with the German Ministry for Post and Telecommunications and the German Space Agency (DARA). He has been delegate to numerous international forums and recently served as the chairman of various European and global committees (ESA International Relations Committee, UNCOPUOS working groups). Kai-Uwe Schrogl has published seven books and more than 100 articles, reports and papers in the fields of space policy and law as well as telecommunications policy. He is Member of the Board of Directors of the International Institute of Space Law, Member of the International Academy of Astronautics (chairing its Commission on policy, economics and law) and the Russian Academy for Cosmonautics as well as member in editorial boards of international journals in the field of space policy and law (Acta Astronautica, Space Policy, Zeitschrift für Luft- und Weltraumrecht, Studies in Space Law/Nijhoff). He holds a doctorate degree in political science, lectures international relations at Tübingen University, Germany (as a Honorarprofessor) and has been a regular guest lecturer i.a. at the International Space University and the Summer Courses of the European Centre for Space Law.

Richard Tremayne-Smith is currently retired. Before his retirement he was the Head of Space Environment and Head of International Relations at the British National Space Centre. From the mid 1960s he was an apprentice at the Royal Aircraft Establishment, Farnborough where his assignments included work in the Department on the UK X3 or Prospero satellite that was later launched on Black Arrow. During the 1970s and 1980s he worked on scientific computer systems leading to European projects on advanced architectures including parallel, AI and knowledge based systems. Through the 1990s until today he worked on a wide range of space related areas including space transportation and propulsion, manned space flight, small satellites, international relations and UN issues, as well as the space environment involving mainly Space debris and near-Earth-objects (NEOs). He provided support to the UK licensing process implementing the provisions of the outer space treaties.

Jean Pierre Swings is currently Professor at the University of Liège, Belgium. Regarding his educational background he obtained an M.A. in Physical Engineering (Space Sciences) in 1965, a Ph.D. in 1969 (Doctorat) and a D.Sc. in 1974 (Agrégation de l'Enseignement Supérieur) at the University of Liege. Jean Pierre's fields of research have included, inter alia, Solar physics, Atomic spectroscopy and Infrared excess objects. His current field of research is in Extragalactic astrophysics (quasars, gravitational lenses, . . .), (Very) large telescopes and instrumentation, Space astrophysics and Solar system exploration. In addition to the over 170 publications on the fields of research listed above and on more general topics Jean Pierre is very much involved in additional activities at several international organizations amongst which are ESA, EAS and ESO.

Jean Claude Worms is currently Head of Unit for Space Sciences – Executive Scientific Secretary of the European Space Sciences Committee (ESSC), ESF's Strategic Board on space research. Amongst many of Jean Claude's achievements are the constitution of a worldwide network of contacts and liaisons among executives in space research; establishment of a successful working interface with the European Commission DG Enterprise; and securing of an observer status in ESA's Ministerial Conferences and EC's FP7 Space Advisory Group. Regarding his educational background Jean Claude obtained a M.Sc. of Applied Physics and a Ph.D. in Physics (astronomy & space science) at the Université Paris 6. His research areas include: radiative transfer in granular media (application to solar system dust); pre-planetary aggregation processes; space debris. Jean Claude is also active in several international institutions, inter alia, such as: Member of the Editorial Board of the International Journal on Nanotechnologies (2003–), Main Scientific Organizer of sessions on solar system small bodies during the COSPAR General Assemblies in 1998 (Nagoya) and 2002 (2nd World Space Congress, Houston), Deputy Organizer during the COSPAR Assemblies in 2000 (Warsaw) and 2004 (Paris), Scientific Organising Committee member during the COSPAR Assemblies in 2006 (Beijing) and 2008 (Montreal). In addition Jean Claude is active member of COSPAR Associate since 1996, EUROSCIENCE since 1998 and ELGRA (European Low-Gravity Association) 1994–2003 and 2006–2007.

Authors and contributors during the conference "Humans in Outer Space – Interdisciplinary Odysseys" on 8–9 October 2007 at Vienna. From left to right. First row: Kai-Uwe Schrogl, Agnieszka Lukaszczyk, Monique van Donzel, Gerda Horneck, Luca Codignola. Second row: Gabriella Cortellessa, Debbora Battaglia, Ulrike Bohlmann, Ulrike Landfester, Frans G. von der Dunk, Olivier Francis, Marcel Egli. Third row: Andreas Happe, Richard Tremayne-Smith, James Muldoon, Jacques Arnould, Pascale Ehrenfreund, Paolo Musso, Jean-Pierre Swings, Gísli Pálsson, Helmut Spitzl, Nicolas Peter, Stephan Lingner, Jean-Claude Worms.

Acknowledgements

The editors would like to thank the following persons and institutions for their valuable contributions in preparing this book. Generous financial support was received by the European Science Foundation (ESF), the European Space Agency (ESA) and the Austrian Federal Ministry for Transport, Innovation and Technology (BMVIT). Particular thanks go to Ms. Silvia Schilgerius from SpringerWienNewYork, who encouraged ESPI to launch a series "Studies in Space Policy" with this book and who, together with her colleagues from all departments of the publishing house, especially to be named Ms. Ursula Szorger, so professionally and kindly accompanied this project. Invaluable support was received by Ms. Blandina Baranes, ESPI's documentalist, who organized the liaison with the publisher and provided additional assistance in various ways. Ms. Minakshi Werner was an efficient proof-reader. And finally, the editors would like to thank Mr. Edi Keck, who participated in the conference and gave a fascinating and inspiring presentation on "Marketing and Branding Space", which, due to technical reasons, could not be included in this volume.

SpringerEngineering

Kai-Uwe Schrogl, Charlotte Mathieu,
Agnieszka Lukaszczyk (eds.)

Threats, Risks, and Sustainability
Answers by Space

2008. approx. 250 pages.
Hardcover approx. **EUR 119,95**
Recommended retail price.
Net-price subject to local VAT.
ISBN 978-3-211-87449-3
Studies in Space Policy
Due March 2009

Development is challenged by, at least until 2050, a strong population growth, more severe environmental strains, growing mobility, and dwindling energy resources. All these factors will lead to serious consequences for humankind. Inadequate agricultural resources, water supply and non renewable energy sources, epidemics, climate change, and natural disasters will further heavily impact human life. European Space Policy Institute (ESPI), sheds a new light on threats, risks and sustainability by combining approaches from various disciplines. It analyzes what could be the contribution of space tools to predict, manage and mitigate those threats. It aims at demonstrating that space is not a niche but has become an overarching tool in solving today's problems.

SpringerWienNewYork

P.O.Box 89, Sachsenplatz 4–6, 1201 Vienna, Austria, Fax +43.1.330 24 26, books@springer.at, **springer.at**
Haberstraße 7, 69126 Heidelberg, Germany, Fax +49.6221.345-4229, SDC-bookorder@springer.com, springer.com
P.O. Box 2485, Secaucus, NJ 07096-2485, USA, Fax +1.201.348-4505, service@springer-ny.com, springer.com
Prices are subject to change without notice. All errors and omissions excepted.

SpringerEngineering

Kai-Uwe Schrogl, Charlotte Mathieu,
Nicolas Peter (eds.)

Yearbook on Space Policy 2006/2007
New Impetus for Europe

2008. XXVI, 330 pages. 49 figures
Hardcover **EUR 119,95**
Recommended retail price.
Net-price subject to local VAT.
ISBN 978-3-211-78922-3
The Yearbook on Space Policy, Series Ed.: ESPI

The Yearbook on Space Policy aims to be the reference publication analysing space policy developments. Each year it presents issues and trends in space policy and the space sector as a whole. Its scope is global and its perspective is European. The Yearbook also links space policy with other policy areas. It highlights specific events and issues, and provides useful insights, data and information on space activities.

The Yearbook on Space Policy is edited by the European Space Policy Institute (ESPI) based in Vienna, Austria. It combines in-house research and contributions of members of the European Space Policy Research and Academic Network (ESPRAN), coordinated by ESPI. The Yearbook is addressed to decision makers in governments and agencies, professionals in industry as well as the service sectors, researchers and scientists and also to the broader public interested in the field.

 SpringerWienNewYork

P.O.Box 89, Sachsenplatz 4–6, 1201 Vienna, Austria, Fax +43.1.330 24 26, books@springer.at, **springer.at**
Haberstraße 7, 69126 Heidelberg, Germany, Fax +49.6221.345-4229, SDC-bookorder@springer.com, springer.com
P.O. Box 2485, Secaucus, NJ 07096-2485, USA, Fax +1.201.348-4505, service@springer-ny.com, springer.com
Prices are subject to change without notice. All errors and omissions excepted.

CPSIA information can be obtained at www.ICGtesting.com
Printed in the USA
LVOW10*1217160314

377605LV00012B/475/P